ERIC TEMPLE BELL

THE LAST PROBLEM

Revised and updated by Underwood Dudley

Mathematical Association of America

The Last Problem

The Last Problem

E. T. BELL

Introduction and Notes by
Underwood Dudley

MAA
SPECTRUM

PUBLISHED BY THE
MATHEMATICAL ASSOCIATION OF AMERICA

This book was updated and revised from the 1961 edition
published by Simon and Schuster, Inc.

Library of Congress Catalog Card Number 90-63128
ISBN 0-88385-451-1

Manufactured in the United States of America

Contents

Introduction

The Last Problem is Eric Temple Bell's last book; he died in December, 1960. When it appeared in 1961 it was not widely noticed, for which there are several possible reasons. One is that prospective reviewers may not have wanted to write reviews that would have contained statements like, "While not to be compared with Bell's monuments, *Men of Mathematics*, *The Development of Mathematics*, and *Mathematics, Queen and Servant of Science*, his latest work nevertheless...." Another is that they may not have known what to make of it. Bell said that the book is the biography of a problem, the famous problem of showing that Pierre Fermat was not mistaken when he wrote in the margin of his copy of Diophantos's *Algebra*, almost 350 years ago, that the equation

$$x^n + y^n = z^n$$

has no solution in positive integers when $n \geqslant 3$. There are not many biographies of problems, and reviewers can be excused for not knowing how to deal with one. The book fits no categories. It is not a book of mathematics. Pages go by without an equation appearing, and in mathematics books you are not told such things as that the ancient Spartans were "as virile as gorillas and as hard (including their heads) as bricks" or that seventeenth-century France "was a cesspool of corruption, yet it survived and became even more corrupt." It is not a history book, either. History books do not contain the nine equations in ten unknowns that come from the cattle problem of Archimedes or proofs that

if p and $2^p - 1$ are prime, then $2^{p-1}(2^p - 1)$ is perfect.

It is not a history of number theory because it includes too much about the history of the western world, and it is not a history of western civilization because its focus is on mathematics. It is too entertaining to be scholarly and contains too much mathematics to be widely popular. It is an unusual book.

But then Eric Temple Bell was an unusual person. He was born in Aberdeen, Scotland in 1883. He said of his early life

My father came of a prominent mercantile family in the City of London; my mother's people were classical scholars for several generations back. My education was by tutors till I entered the Bedford Modern School, where the late E. M. Langley converted me to mathematics.

His father was James Bell and his mother was Jane Lindsay-Lyall; Edward Mann Langley was a mathematician of no mean ability who Bell was fortunate to encounter in secondary school.

To escape being shoved into Woolwich or the Indian Civil Service, I left England and came on my own at nineteen to California, where I matriculated at Stanford University. There (in 1902) the free elective system was still in force. I took only mathematics. Langley had done such a good job of teaching that in two years I covered all the mathematics offered.

Bell does not mention that he had a remarkable talent for mathematics. In 1904 he became a Stanford B.A.

The rest of my education was one year at the University of Washington and another at Columbia University where (1912) I took my Ph.D. in mathematics.

His second bachelor's degree was granted in 1908. Why a second degree? Why did Bell come to the United States in the first place? How did he live? Did he ever see his parents again? How did he get a Ph.D. degree in only one year? There are many questions, and since Bell has had no biographer they will probably go forever answered.

In the meantime, I had worked as a ranch hand, a mule skinner in Nevada, a surveyor, partner in a telephone company in San Francisco and had gone broke in the great earthquake and fire of 1906. After clearing up with my partners our obligations, I quit business for good and returned to mathematics and, as a recreation, writing.

He married in 1910 and was the father of one son. He listed his recreations as gardening and cats.

Bell joined the faculty of the University of Washington in 1912 and moved to the California Institute of Technology in 1926 where he remained until his retirement in 1953. He was amazingly prolific, with more than 250 mathematical papers to his credit, mainly in number theory, concentrating on Diophantine equations. His name is not to be found in the histories of twentieth-century mathematics as one of the major contributors to number theory but then few are, and those few needed the support of the many whose contributions have faded away. His contemporaries thought very highly of him, electing him a vice-president of the American Mathematical Society in 1926, president of the Mathematical Association of America in 1931–33, vice-president of the mathematical section of the American Association for the Advancement of Science in 1939, and member of the National Academy of Sciences. Besides these services to the profession, besides producing several papers every year, and besides the duties attendant on being paid a salary by a university, he was turning out books intended for a general audience. *Men of Mathematics* (1937), *The Development of Mathematics* (1940), and *Mathematics, Queen and Servant of Science* (1951) are his major works, but there were others: *Handmaiden of the Sciences* (1931), *Numerology* (1933), *The Search for Truth* (1935), *Man and his Lifebelts* (1938), *The Magic of Numbers* (1946). One wonders when the garden and the cats got any attention.

But that is not everything. He was turning out great quantities of science fiction using the pseudonym John Taine. *The Purple Sapphire* (1924), *Quayle's Invention*, (1926), *The Gold Tooth* (1927), *Green Fire* (1927): the titles of the novels go on and on, more than a dozen of them, not ending until *The Crystal Horde* (1952) and *G. O. G. 666* (1952). He thought highly enough of one, *Before the Dawn*, to publish it under his own name. Anyone who publishes novel after novel must be good at his craft, and a contemporary critic said that his books were "noted for their violence and exciting plots." *Green Fire* —about a mad scientist bent on destroying the world—was even turned into a play, set in 1990, that had more than one production. ("Alan's so cautious! He won't drive over two hundred even at night, with the speedway tunnels practically empty. You'd think he was living back in nineteen-forty.") Bell also produced a volume of poetry. He was a marvel of energy and application. There is no mathematician like him today.

4

T. A. A. Broadbent's assessment of Bell's work is accurate:

His style is clear and exuberant, his opinions, whether we agree with them or not, are expressed forcefully, often with humor and a little gentle malice. He was no uncritical hero-worshipper, being as quick to mark the opportunity lost as the ground gained, so that from his books we get a vision of mathematics as a high activity of the questing human mind, often fallible, but always pressing on in the neverending search for mathematical truth.

Yes, that describes *The Last Problem* exactly.

Sources: *Dictionary of American Biography*, supplement 6, 1980; *Nature*, February 11, 1961; *The New York Times*, December 22, 1960; *Twentieth Century Authors*, first supplement, 1955; *Who's Who*, 1959; *Who Was Who in America*, volume 4, 1968.

Underwood Dudley

1

Prospectus: Unfinished Business

This is a double biography. It has two heroes: a problem, of which hints can be traced back to the Babylonia of about 2000 B.C., and a man, Pierre Fermat, 1601–1665, who in 1637 set the problem in its present form. The idea of such a biography was inspired—if that is the right word—by the atomic bomb and its successors: the hydrogen bomb, the U-bomb, and from there on out as far as a nuclear physics can go before the end, if there is to be an end.

Suppose that our atomic age is to end in total disaster. Civilization will be wiped out, and with it all but a scattered handful of human beings too deeply diseased to start the long climb up from primitivism. What problems that our race has struggled for centuries to solve will still be open when the darkness comes down? A philosopher might suggest several, such as the nature of "reality"; a moralist could propose the problem of good and evil; a sociologist might ask how to abolish poverty and war; and so on. But problems such as these have not yet been stated sharply enough for a moderately critical onlooker to understand precisely what they are about. When the proposers disagree among themselves on the meanings of their problems, realists may be pardoned for suspecting that some are pseudo-problems incapable of solution. So we shall leave them aside and look for others on an understandable level, in the hope of finding one or two that make simple sense and conceivably admit definite answers, although we have not found them after hundreds of years.

Where shall we look for such problems? Current science seems to offer many—the nature of life, for instance, or the ultimate constitution of matter and radiation. But most of these again are either ambiguous or too broad for exact statement. The first may not even make sense. So we shall have to be content with something that anyone with an elementary-school education can understand, no matter how trivial it may seem at a first glance. Here a promising lead is the recorded history of *the most elementary mathematics, particularly arithmetic*.

The broad outlines of the relevant history, from ancient Sumeria, Babylonia, and the Egypt of about 2000 B.C., to A.D. 1958, are reasonably clear. We may profitably explore these. Our candidates for outlasting humanity should not only be easily understandable by any person of normal intelligence with an ordinary education; for at least a century they should also have withstood the strongest attacks of some of the greatest mathematicians in history. The time limit is set to ensure that the problems are really hard, however easy or trivial they may seem to those who have not seriously tried to settle them, or who may be unacquainted with the part these simple arithmetical problems have played, and continue to play, in the long development of mathematics, both pure and applied.

Two problems present themselves immediately. The older one dates from the fourth century B.C., and is Greek. Of all the mathematical questions left by the Greeks, this is the only one that is still unanswered. Though the Greeks did not state it explicitly, it is at once suggested by some of their earliest discoveries. It seems approachable and may be solved before the end. Although it is not the main problem to be discussed, I include it because it was the source of much ingenious but inconclusive work from the seventeenth century to the 1950s, and also because it has a curious connection, discovered only in 1938, with the second and possibly harder question. It concerns a peculiar property of certain common whole numbers, which I shall describe here ahead of its history. Those who wish to get on at once to the history may pass to the next chapter.

The sequence of *natural numbers*, or the *positive integers*, $1, 2, 3, 4, 5, 6, \ldots$ is the basis of arithmetic, both elementary and higher. Each of these numbers after 1 is exactly divisible without remainder by at least two numbers in the sequence; thus 2 is divisible by 1 and 2, or $2 = 1 \cdot 2$ (the dot is read "times," thus 1 times 2); 3 is divisible by 1 and 3, $3 = 1 \cdot 3$; 4 is divisible by 1, 2, and 4, $4 = 1 \cdot 4 = 2 \cdot 2$; 5 is divisible by 1 and 5, $5 = 1 \cdot 5$; 6 is divisible by 1, 2, 3, and 6, and so on. The numbers that divide a given number exactly (without remainder) are called the *divisors* of the number. If 2 is a divisor of a number, the number is called *even*, otherwise *odd*. The even numbers are $2, 4, 6, 8 \ldots$, the odd are $1, 3, 5, 7, 9 \ldots$. Those divisors of a given number that are *less than the number itself* are called the *aliquot parts* of the number. For example, the aliquot parts of 6 are $1, 2, 3$. Following the Pythagoreans in the sixth century B.C., and Euclid in the third and fourth century B.C., we note that the sum, $1 + 2 + 3$, of the aliquot parts of 6 is equal to 6. A number which is equal to the sum of its aliquot parts is called *perfect*. By trial, we find that *the next* perfect number after 6 is 28; all the aliquot parts of 28 are 1, 2, 4, 7, 14, and their sum is 28. With sufficient persistence the reader may verify that *the next* perfect number after 28 is 496, *the next* 8128, *the next* 130816, *the next* 2096128, *the next* 33550336, *the next* 8589869056, ... Notice that these first seven perfect numbers end in 6 or 8, so all are even. All the perfect numbers so far discovered are even and all end in 6 or 8. The first part of the problem of perfect numbers is to *prove or disprove that an odd perfect number exists*. This is included in the second and probably much harder part: *find all perfect numbers by means other than trial*.

About 300 B.C., Euclid stated and proved the *sufficient* form of an even perfect number (see Chapter 6), and Euler in the eighteenth century proved that Euclid's form is also *necessary*, but neither of these first-rank mathematicians gave any method much better than trial for finding the successive even perfect numbers. The problem is deep. We shall have to know much more than we do about prime numbers before a decisive attack is feasible. A *prime number*, or briefly a *prime*, is a *number greater than 1* having only 1 and itself as divisors; for example, 2, 3, 5, 7, 11, 13, 17, 19, 23, 29, are the first ten primes. Note that 1 is not counted as a prime; the only even prime is 2. As will appear when we come to Euclid, the problem of finding even perfect numbers is equivalent to that of discovering primes of a certain kind. One of the modern calculating machines invented for use in the Second World War was released when not engaged in military work for a few hours now and then to explore the sequence 1, 2, 3, 4, 5, ... for perfect numbers. The calculations this machine did in a matter of hours or a few days were far beyond the capacity of the entire human race toiling twenty-four hours a day continuously for months or years. But the machine did not solve the ancient Greek problem of perfect numbers. Machines cannot think. I shall have considerably more to say about these numbers in later chapters.

The second candidate for the possible distinction of outlasting the human race is a French problem dating from 1637. Fermat is responsible for this:

Prove or disprove that if n is a number greater than 2, there are no numbers a, b, c such that

$$(\mathbf{F}) \qquad\qquad a^n + b^n = c^n.$$

"Number" here means positive integer as already defined—common (natural) whole number. For those who have forgotten how to read algebraic symbolism,[1] a^n means $a \cdot a$, where there are n a's; a^n is called the nth *power* of a; for example, $a^3 = a \cdot a \cdot a$, the third power of a, $5^4 = 5 \cdot 5 \cdot 5 \cdot 5 = 625$ and so on. Similarly, for b^n, c^n; so the problem may be restated thus: to prove or disprove that if n is greater than 2, the sum of the nth powers of two positive integers is never equal to the nth power of a positive integer.

The exception $n = 2$ is necessary since, for example, $3^2 + 4^2 = 5^2$, that is, $9 + 16 = 25$, and, as the Babylonians and

---

[1]Possibly I should apologize for recalling the following details of elementary algebraic symbolism that older readers learned in school. But experience with highly educated persons has suggested that an explanation might be helpful. Even specialists in the humanities sometimes confuse $E = mc^2$ with $E = 2mc$; the first is Einstein's equation connecting energy, E, with mass, m; c, in the appropriate units, denotes the speed of light. If the second equation were true, which it is not, there would be no atomic bombs and no atomic age. Anyone who has ever taught beginners in algebra will remember the difficulty they had in distinguishing c^2 from $2c$. That "misplaced" 2 is the difference between heaven and hell. So beginners may be right in asserting that algebra is the very devil.

Plato knew, there are actually an infinity of positive-integer solutions of $a^2 + b^2 = c^2$. Fermat asserted that his equation (**F**) *is impossible in numbers a, b, c, n if n exceeds* 2. This assertion is known as "Fermat's Last Theorem," or "The Great Fermat Theorem." He said that he had a proof.

Why have mathematicians bothered with Fermat's unsubstantiated claim? Possibly because it is a challenge to the powerful methods of the mathematics developed since Fermat's seventeenth century, and pride in craftsmanship obligates the mathematicians of one generation to dispose of the unfinished business of their predecessors. More objectively, numerous unsuccessful attempts to dispose of Fermat have resulted in deep theories with many applications to both pure and applied mathematics, and from there to science in general. Without the initial stimulus of Fermat's baffling assertion, none of these useful things might have been invented. But possible utility has played only a very minor part compared to sheer curiosity.

It may be of interest to say how I came to write this account of what led up to Fermat, and the conditions under which he and his predecessors made their simple but profound discoveries. Having been long acquainted with the mathematics concerned, I became interested in its creators as human beings and men of their times; and when the opportunity came I tried to find out something about them. Sometimes there was nothing or but very little. The lives of many are almost unknown, or compressed to a dry sentence or two in the standard histories of mathematics. The Babylonian mathematicians have left not even their names; a few of the astronomers have. But much is known about the civilizations in which all these mathematicians did their enduring work. Some of this may suggest what the lives of the men concerned may have been like. The truism that a man is a product of his times suggests that we look at the man's times when he himself is not in plain view. If most of his contemporaries were brutal and callous according to the morals we profess, it is unlikely that he invariably was considerate to his fellow men. The most to be expected of him is a protective shell of indifference without which he could hardly have got on with his work. We shall see instances of this, especially in Euclid's Alexandria of the third and fourth centuries B.C., and Fermat's seventeenth-century France. Funeral orations, obituaries and official biographies of scientific men either take the shell for granted and say nothing about it, or give it a thick coat of whitewash. So it comes as a shock to find that a great man who seemed to be above the barbarism of his times was after all in some respects no higher than the degradation, the corruption, the slavery, and the cruelty that made it possible for him to live and work in ease and security. But the shock is unreasonable. We need only to look about us. The pattern persists.

Until we scan the record we might imagine that peace is a necessary condition for the creation of lasting mathematics. It was not so in Euclid's and Fermat's times. Much of Fermat's best work was done while one of the most savage wars in history raged all about him. Yet he never alludes to it in his correspondence. Even the Alexandria that fostered the golden age of Greek mathematics owed its foundation to the wars of Alexander the Great; and while Euclid and his colleagues were serenely mathematicizing, recurrent wars increased the wealth and prestige of that great city, Alexandria, and the oppressed peasantry fled to the swamps of the Nile because they could "stand no more." Again the pattern persists into our own times. The warp is squalor, grinding labor, poverty to starvation, crude bestiality, inhuman (or human?) brutality, and the woof, polite refinement, ease, luxury, knowledge, learning, and science. Of course there are gray threads between the black and white, but they are rather rare.

One of the following chapters gives an account of the life and times of Fermat, founder of the theory of numbers and one of the great mathematicians of history. Not a mathematician by profession, Fermat never held any academic position. He approached mathematics as an amateur and attained the first rank. He is the only mathematical amateur in history of whom the last is true.

Whatever of a man's "life" is worth remembering may extend from thousands of years before he was born to centuries after he is dead. It is so with Fermat. To understand his work we shall have to go far back to its beginnings in Babylonia, and from there follow down the tenuous thread to the seventeenth century in France. *Only those items out of all the incredibly rich mathematical history of about 3700 years having a direct bearing in Fermat's discoveries in the theory of numbers will be noted in more than brief and passing detail.* We shall observe what kinds of societies and individuals contributed to this amateur mathematician's decisive achievements in one of the most difficult—though apparently the simplest—departments of mathematics.

If some of what I have included may seem remote from mathematics, my reason is that even mathematicians have been interested in the more human side of their fellows. The geometer Guillaume L'hôpital, for example, asked about Newton, "Does he eat, drink and sleep like other men? I represent him to myself as a celestial genius, *entirely disengaged from matter.*" Many of the people we shall meet ate and drank well, and some were up to their chins in the muck of material things.

Perhaps that is why they and their civilizations produced lasting mathematics. The only man we shall encounter in a cloister is Father Marin Mersenne, and he was no insipid saint. To make his otherwise rather drab existence interesting he tempered austerity with good fare and scholarly politics—stirring up bitter disputes between his intellectual friends. Mersenne, incidentally, is one of several mathematicians mentioned only in passing, if at all, in the shorter histories of mathematics. In connection with Fermat, however, he is important. That shifty but on the whole not unlikable scoundrel, Sir Kenelm Digby, is another of these lesser figures who counted in Fermat's life. Of quite a different stature was that celebrated prodigy, John Wallis, a pygmy next to Fermat, who had the effrontery to condescend to the great Frenchman. Wallis never understood what Fermat was talking about, but as an irritant he had an important part in Fermat's mathematical development. Others, now all but forgotten, survive as far as they do chiefly because they irritated Fermat to the point of taking up their challenges. I have given short sketches of the lives of such minor characters where they are of some independent interest.

Fermat's greatest work was in the theory of numbers, of which, as noted, he was the founder as it is developed today. The theory is not concerned with computation. It seeks general properties of classes of numbers. To take a trivial example—it goes back to Nicomachus of Gerasa and to the Alexandria of the first A.D.—what can we say about the even numbers $2, 4, 6, 8, \ldots$ in relation to the odd numbers $1, 3, 5, 7, \ldots$? Among other things, any even number is the sum of two odd numbers: $2 = 1 + 1$; $4 = 1 + 3$; $6 = 1 + 5 = 3 + 3$; $8 = 1 + 7 = 3 + 5$, and so on. What about the odd numbers? Each after 1 is the sum of an even and an odd number. These general and, to us, trivial properties of even and odd numbers are verified by trial and can be easily proved. They could be the first discoveries an amateur might make. Going a little farther, the encouraged amateur might ask how the even numbers are related to the primes. He could find by trial that $4 = 2 + 2$; $6 = 3 + 3$; $8 = 3 + 5$; $10 = 5 + 5 = 3 + 7$. Continuing thus he might be bold enough after about 40,000 trials (the actual limit in the older work) to conjecture that *every* even number greater than 4 is a sum of two odd primes. But induction from special cases not only proves nothing in the theory of numbers but may be disastrous—disastrous because a wrong guess might entail many a wasted lifetime and mislead others into false assertions. Statements about numbers can be true in a billion instances and false in the billion and first. It is easy to construct such statements. As for the question about even numbers and primes, nobody knows whether or not every even number greater than 4 is a sum of two primes, in spite of numerous efforts to settle the question since it was first asked in 1742 by Christian Goldbach. The best so far proved (by I. M. Vinogradov in 1937) is that every "sufficiently large" odd number is a sum of three old primes. The "sufficiently large" could be made precise by mere calculation with modern machines if it were worth while. Vinogradov's proof is by no means elementary or even easy. Failing to settle Goldbach, our hopeful amateur might ask, how are *all* the numbers $2, 3, 4, 5, 6 \ldots$ related to the primes? Here, with reasonable luck, he might rediscover that any number greater than 1 is either a prime or can be made up by multiplying primes, and essentially uniquely. For example, $10 = 2 \cdot 5$; $12 = 2 \cdot 2 \cdot 3$; $123 = 3 \cdot 41$. Although not difficult, the complete proof might well baffle an amateur. As we shall see, Gauss (1777–1855) first proved this completely about 1800, but he was no amateur.

15

The examples just given illustrate L. E. Dickson's statement in the preface of Volume 1 of his classic *History of the Theory of Numbers*:[2]

The theory of numbers is especially entitled to a separate history on account of the great interest which has been taken in it continuously through the centuries from the time of Pythagoras [about the sixth century B.C.], an interest shared on the one extreme by nearly every noted mathematician and on the other extreme by numerous amateurs attracted by no other part of mathematics.

Again, "perfect numbers have engaged the attention of arithmeticians of every century of the Christian era. It was while investigating them that Fermat discovered the theorem which bears his name in elementary texts [stated here in Chapter 12] and which forms the basis of a large part of the theory of numbers." Probably it was ordinary amateurs who first discovered the smallest perfect numbers—6, 28, 496—but it took the penetration of a great mathematician—Euclid—to get anything of significance out of the search for perfect numbers.

The experience of amateurs and professionals alike shows that whoever hopes to find anything of interest about numbers will do well to experiment with the numbers themselves. It is not necessary to preserve the computations and the guesses that suggested the final result and its proof. In fact, it seems to have been a point of false pride for arithmeticians to cover up the tracks by which they reached their goals—as likely as not unforeseen when they started. Fermat and his contemporaries

---

[2]Three volumes totaling 1637 pages, Carnegie Institution of Washington, 1919, 1920, 1923.

indulged freely in this exasperating mischief, so that often we are ignorant whether or not they had proved what they claimed. Though the resulting mystification may have flattered their vanity and increased their prestige, it did the theory of numbers no particular good. In the end the mystifiers robbed themselves of the fame they coveted, and history credits them only with the rash guesses—"conjectures" is the polite word—they may or may not have proved. Sometimes they guessed wrong, and were shown up long after they should have been sleeping peacefully in their graves.

Almost anyone, after a little experimenting, can make a plausible conjecture about numbers, but only a foolhardy optimist today publishes his guesses. Proof or nothing is the rule for reputable mathematicians. It was not so in Fermat's day. Gauss, usually bracketed with Archimedes and Newton as one of the three greatest mathematicians in history, and an arithmetician of the highest rank, deprecated unsubstantiated guessing. A friend had asked him why he did not compete for the prize offered in 1816 by the French Academy of Sciences for a proof (or disproof) of Fermat's Last Theorem. "I confess," he replied, "that Fermat's Theorem as an isolated proposition has very little interest for me, because I could easily lay down a multitude of such propositions, which one could neither prove nor dispose of." Though he never said so explicitly, he seems to have doubted that Fermat had proved his theorem.

Before passing on to the slow centuries of evolution that finally produced Fermat and his work, I transcribe two tributes to the theory of numbers, to suggest why "the higher arithmetic" has attracted amateurs and professionals alike for more than twenty centuries. The first (1847) is from Gauss, as translated by H. J. S. Smith, himself a great arithmetician; the second (1859) is from Smith.

17

The higher arithmetic presents us with an inexhaustable store of interesting truths—of truths too, which are not isolated, but stand in a close internal connexion, and between which, as our knowledge increases, we are continually discovering new and sometimes wholly unexpected ties. A great part of its theories derives an additional charm from the peculiarity that important propositions, with the imprint of simplicity upon them, are often easily discoverable by induction, and yet are of so profound a character that we cannot find their demonstration till after many vain attempts; and even then, when we do succeed, it is often by some tedious and artificial process, while the simpler methods may long remain concealed.

The Theory of Numbers has acquired a great and increasing claim to the attention of mathematicians. It is equally remarkable for the number and importance of its results, for the precision and rigourousness of its demonstrations, for the variety of its methods, for the intimate relations between truths apparently isolated which it sometimes discloses, and for the numerous applications of which it is susceptible in other parts of analysis.

To forewarn the reader what to expect, and what not to expect, I offer two modifications of Alexander Pope's "Fools rush in where angels fear to tread":
  (1) Mathematicians rush in where historians fear to tread;
  (2) Historians rush in where mathematicians fear to tread.
Finally, before getting on with the job, I transcribe an evaluation of "scholarship" by a noted Egyptologist, Flinders Petrie: "When an author collects together the opinions of as many others as he can and fills half of every page with footnotes, this is known as 'scholarship.'" The very few footnotes in this book that need be noticed relate to (2) and are mildly technical. The others can be skipped.

2

The Far Beginnings:
Babylon and Egypt

1. Babylon

Only seldom can a scientific or mathematical question of living interest be traced back without guesswork to an origin thousands of years ago. For the problem of primary interest here, the line of descent of the theory of numbers from the Babylonia of about 2000 B.C. to Fermat in the seventeenth century is direct, except for a gap to be filled in the period immediately following the fall of Babylon. (Cyrus destroyed Babylon in 539 B.C.) Reasonable extrapolation from both ends of the gap suggests what is missing and probably will be discovered in further archaeological research. From Fermat to the present the line is unmistakably clear; and historical research has shown that the theory of numbers in its scientific, unmystical form had begun in Babylonia well over three thousand years before Fermat was born.

Fermat was directly inspired by Diophantus of Alexandria (dates uncertain; see Chapter 7), the last of the major Greek mathematicians–if indeed he was a Greek. But what peoples preceded Diophantus, and who first hinted at the kind of problem that led Fermat to his still unproved Last Theorem? If any people did, it was the Babylonians of the second millennium B.C., and possibly, but to a much lesser extent, if any, the approximately contemporary Egyptians.

There is no documentary evidence that Diophantus ever heard of the Babylonian beginnings of the theory of numbers. But it is incredible that all the tradition of this suggestive pioneering work should have perished, though even when Diophantus lived, the line of descent may have been forgotten beyond recovery. Probably several centuries separated the Greek from his Babylonian precursors. We can only conjecture that some of the ancient tradition survived the ultimate fall of Babylon, and it is no less than a reasonable guess that the Greek predecessors of the individualistic Diophantus were aware of the great work of the Babylonians and profited by it. As for Diophantus himself, some of his problems might have been adapted from Babylonian cuneiform tablets. Unless we grant this much, we must believe in unlikely mathematical coincidences and improbable historical miracles. We shall therefore assume that the Babylonians influenced the Greeks, but with the reservation emphasized by L. E. Dickson in his exhaustive *History of the Theory of Numbers*:[1]

Again, conventional histories take for granted that each fact has been discovered by a natural series of deductions from earlier facts and devote considerable space in the attempt to trace the sequence. But men experienced in research know that at least the germs of many important results are discovered by a sudden and

---

[1] Especially Vol. 2, *Diophantine Analysis* (Washington, 1920).

mysterious intuition, perhaps the result of subconscious mental effort, even though such intuitions have to be subjected later to the sorting process of the critical faculties.

Before proceeding to Babylon, we shall pause very briefly in Sumeria, just north of the Persian Gulf. The Sumerians were a non-Semitic, non-Indo-European people; their neighbors to the north were the Semitic Akkadians. The Sumerian golden age lasted from the beginning of the fourth millenium to the end of the third millennium B.C., about 2000 years. They either invented or adopted the cuneiform script in which Babylonian mathematics, among other things, is recorded. They also are credited with advances in civil and hydraulic engineering, making possible a vast network of irrigation canals on the Mesopotamian plain, the draining of marshes, and the control of the flood waters of the two great rivers, the Tigris and Euphrates. Without these canals, agriculture and horticulture could hardly have developed as they did.

The evidence of the Sumerians' numerical ability is in the profusion of cuneiform tablets concerning commercial matters —bills, receipts, notes, bookkeeping, computation of interest on loans, weights and measures. The practical utility of all these things is obvious; so far no trace of anything more advanced has been discovered. The sexagesimal system of numerical notation (counting by 60s instead of by 10s), with a slight admixture of the decimal system (by 10s), which the Babylonians exploited in both mathematics and astronomy, goes back to the Sumerians. So also for much law and religion.

The Sumerians developed religious and spiritual concepts together with a reasonably well integrated pantheon which influenced profoundly all the peoples of the Near East, including the

Hebrews and the Greeks. Moreover, by way of Judaism, Christianity, and Mohammedanism, not a few of these spiritual and religious concepts have permeated the modern civilized world.[2]

Again:

The Sumerians produced a vast and highly developed literature, largely poetic in character, consisting of epics and myths, hymns and lamentations, proverbs and "words of wisdom." These compositions are inscribed in cuneiform script on clay tablets which date largely from about 2000 B.C. In the course of the past hundred years, approximately three thousand such literary pieces have been excavated in the mounds of ancient Sumer.[2]

Somewhere in the third millennium B.C., an unknown mechanical and agricultural genius made an invention that was as revolutionary for the Sumerian economy and way of life as the atomic age promises to be for our own. From a distance of nearly five thousand years this shattering invention is more likely to provoke laughter than to inspire awe and respect. Let us hope that if any of our descendants survive till the year 6900 or thereabouts, they will be able to look back and see our invention of the atomic bomb as a ludicrous incident in the history of a primitive civilization, just as we look back and laugh at the Sumerian equivalent of our bomb: the pickax.[3]

The magnificent, if frequently brutal, civilizations of Babylon and Egypt have left traces of their vanished splendors in their imperishable mathematics. Babylon made by far the greater

---

[2] S. N. Kramer, "Sumerian Mythology," American Philosophical Society, *Memoirs*, XXI (1944), 7.

[3] The poem commemorating "The creation of the pickax" is translated by Kramer on pages 52–53 of the paper just cited.

contribution, especially to computation and the theory of numbers, in only the latter of which we are interested here. This however may be the residue of a material accident. The towers and walls of Babylon collapsed many centuries ago into shapeless mounds of crumbled clay, while Egypt's massive temples and pyramids of worked stone, though worn by wind, weather, and pillage, have lasted to our own time. The Babylonians built in sun-dried brick. For reasons indicated presently they inscribed their mathematical lore on smooth, resistant, kiln-baked tablets of clay immune to ordinary corruption. When the tiered temples and libraries of sun-dried brick lapsed into ruin the hardened tablets slowly sank into soft mud or wind-blown dust. Centuries later hundreds of thousands were exhumed, some shattered or badly chipped, but many more almost as sound and as legible as the day they were drawn from the kilns. This happy fortune cannot be credited to any acute foresight on the part of the Babylonians. Their rich alluvial plain, crisscrossed by canals to control and distribute the waters of its two life-giving rivers, the Euphrates and the Tigris, grew no plant suggesting a natural paper. But the water-logged soil did supply an unlimited store of common mud and choice clay for the slave-driving engineers' millions of bricks and the diligent scribes' hundreds of thousands of well-baked tablets. Then, after thousands of years, the preserved tablets were unearthed, exposed to air and weather, and rapidly began disintegrating.

The seemingly more fortunate Egyptians had a perennially renewed stock of the graceful papyrus sedge in the marshes of the Nile, and from this they macerated their paper—papyrus, as we call it. Papyrus in an arid climate is about as durable as our best grade of rag paper. Whatever the Egyptian priests knew of science and mathematics seems to have been entrusted mostly to papyrus; there is very little of scientific interest on their monuments and frescoes. In the dry climate of

Egypt, if buried in the sand or shut up in a tomb, papyrus is practically immortal. The huge blocks of stone the tyrants transported long distances at the cost of generations of grinding slavery were reserved for the inscription of religious observances and the boasts of egoistic conquerors, happily interspersed however with numerous vivid portrayals of the daily life of the working people. While we may be grateful for these artistic masterpieces, we can only regret that their lush environment seduced the Egyptians into the fabrication of a perishable paper. They also wrote on leather.

The contrast between the surviving records of these two dead civilizations accentuates our own plight. All of our living science is recorded on paper far less resistant than papyrus. One major war in the modern fashion—with A-bombs, H-bombs, U-bombs, disease germs, and the rest—could obliterate it all in a month, and sweep our scientific and technological civilization back to the devout and untroubled ignorance of the Christian Dark Ages.

Mathematics was made by men. This blunt truism denies Plato's celestial creed, still upheld by some schools of philosophy, that mathematics existed in heaven for eons before mere human beings were imagined by the Divine Mind to perceive the mystical harmonies of arithmetical addition, subtraction, multiplication, division, and common fractions. The Divine Mind seems, however, to have been confused, if not baffled, by square roots. It took Plato's predecessors a long time to find out that the length of the diagonal of a square with each side one unit in length, which equals the square root of two, is not expressible as a common fraction, as it should have been in celestial arithmetic.

Accepting the human origin of the theory of numbers we shall see first what kinds of peoples were responsible for the beginnings of the theory. I shall report incidentally some of what Herodotus (born 484 B.C.) had to say of the Babylonians and the Egyptians (when we come to them) as he observed them with his own eyes. Egypt had petrified mathematically long before he saw it, and the greatest age of Babylonian mathematics was past when he toured Mesopotamia. Nevertheless, what he described may give us some idea of the inquisitive and ingenious peoples who took the first recorded steps toward our current theory of numbers.

It used to be a mark of scholarly (or pedantic) distinction to dismiss Herodotus as a gullible Greek gadabout eager to believe every old wives' tale the priests foisted off on him. He has been variously called the Father of History and the Father of Lies. Modern historical research has shown that sometimes he did believe too much of the alleged history his sly friends and priests—who were also the historians and scientists of Egypt and Babylon—palmed off on him as truth, but so far it has not been proved that he was untruthful about what he says he actually saw. An occasional observation vindicates him now, as it did not before modern science caught up with him. The significant item of boiled drinking water will be noted later. Even more striking are his reports on date culture and the part of certain highly specialized wasps in the fertilization of figs. What he records even now makes possible thriving agricultural industries today in Southern California. As for mathematics, his observations, though tantalizingly scant, are acute.

Herodotus and others portray the Babylonians and the Egyptians as decent peoples, except in war, when both, especially the Babylonians, were savagely cruel and sadistic. To surpass their wholesale inhumanities we have to come down to Fermat's seventeenth century and our own European 1930s and 1940s. As the Babylonians were more directly responsible than the Egyptians for the first steps toward a theory of

25

numbers, and as Herodotus seems to favor them over the Egyptians, I shall report first some of what he says about Babylonia and its inhabitants.

His description of the Babylonian countryside might fit the rich Imperial Valley and the subtropical Coachella Valley of Southern California today. The hostile, torrid climate was subdued by a network of canals spreading the waters of the Tigris and the Euphrates over the whole country. Without this intricate irrigation system horticulture and agriculture would have been impossible. There was no attempt, according to Herodotus, to grow the vine, the olive. Lacking olives, the Babylonians had no soap in the sense of the ancient Greeks, and instead of olive oil they used oil of sesame. The vast plain was dotted with palm trees supplying an abundance of food, wine, and honey. Herodotus' account of date culture might refer to the Indio district near that fabulous Mecca of Hollywood, Palm Springs. When he describes the luxuriance of the cereal crops he doubts whether he will be believed. The harvests were two-hundredfold, or even three-hundredfold in the best years. The heads of "wheat" were "easily four fingers broad." He cannot have been talking of what we call wheat, although Californians or Iowans might say his report errs on the side of modesty. And as for millet and sesame, he will not speak, although he knew, he says, how tall they grow. For he is well aware that even what he had said of corn ("wheat") is utterly disbelieved by those who have never visited Babylonia —or California or Iowa.

Herodotus has a suggestive note on the Babylonians' transportation by water. They navigated the Euphrates downstream from Armenia in "round boats shaped like a shield," made of hides stretched tight over willow frames. Instead of attempting the lung-bursting upstream return voyage from Babylon, the boatmen stripped off the hides, sold the willow frames, loaded the hides on asses, and drove their beasts back to Armenia. In the Moscow Papyrus there is a statement giving the area of a hemisphere, not accepted by all scholars. If it is accepted for what it purports to be, it may have been suggested by the "round boats," as a rough measure of capacity.

Always interested in the life of the peoples he visited, Herodotus tells us what the Babylonians ate. In addition to cereals and dates, dried fish was a staple of the diet (some classes ate nothing else), either mashed into a paste and eaten raw, or baked. From other accounts it appears that the Babylonians appreciated the dubious virtues of wine. In fact the somewhat confused Biblical story of the fall of Babylon says Belshazzar and his guests were drunk when the Persians surprised them at their orgiastic banquet.

When Herodotus describes the City of Babylon he lets himself go. A difficulty in understanding his account is that we may not know the present equivalents of his "talents," "cubits" and "furlongs." The core of the City of Babylon was one wall within another. After the destruction of Nineveh (612 B.C.) the city became the gigantic fortress capital of Assyria. It loomed up like a mountain range over the vast plain where its ruins can still be traced. This "queenliest city of antiquity" was laid out in the form of a square, fifteen miles to the side. Around the outer walls ran a deep and wide moat brimful of water. The walls, ninety feet high, were so broad that a four-horse chariot could be driven between the sentry boxes along the top. The baked bricks for the construction of this mightiest fortification in history were pressed by hundreds of thousands of slaves from the clay laboriously excavated in digging the moat. Here we may note a mathematical detail. The Babylonian tablets record many calculations concerning the amount of dirt a laborer could move in a day and his keep while on the job. Such economic arithmetic undoubtedly was partly responsible for the Babylonians' preoccupation with mensuration and numerical calculation.

The city's outer wall was pierced by a hundred bronze gates with bronze posts and bronze lintels. Another detail brings us abreast of modern international intrigues to acquire the lucrative petroleum of the Near East, especially Iraq and Iran. Every thirtieth tier of bricks in the great wall was covered by a layer of wattled reeds cemented with bituman—the crudest of crude oil in our language. This precious pungent stuff was brought an eight days' journey to Babylon from Is (Hit or Ait today), where ran a little river with "many big gouts of bitumen."

The city's first and strongest line of defense was this outer wall. An inner wall, less massive, opposed the second and final obstacle to an invader. Between the two walls was sufficient acreage for the cultivation of millet and other cereals to sustain the city's population throughout a reasonably prolonged siege. Surely if ever any city was impregnable to assault from without, Babylon was. It fell at last, because then as now neither bronze nor high brass and brute strength is a match for brains.

Babylon in the time of Herodotus was laid out in a checkerboard pattern of straight streets as in many American cities today. The Euphrates bisected the city. One system of streets ran at right angles to the river, and each street on either side of the river was blocked by a bronze gate. The river was banked by walls of baked and glazed brick. The city had many three and four-story houses. All of the important buildings, such as the king's palace and the temples, were surrounded by massive high walls.

The report of the temple of Bel (Biblical Baal) particularly impressed Herodotus. Protected by a thick square wall a quarter of a mile to the side, with gates of bronze, the temple court enclosed a solid tower (ziggurat) an eighth of a mile square.

This huge mass of hand-made brick was a succession of eight diminishing towers piled one on top of the other to the highest, with an outer staircase round all the towers. Halfway up there were resting places for the short-winded. (Children's illustrated Bibles used to picture the Tower of Babel as a ziggurat.) Crowning it all was a "large well-covered couch and a golden table." No image presided over this shrine, where a Babylonian virgin lay alone at night awaiting the advent of her unknown god and lover. Herodotus does not say whether the god ever arrived, but it appears that less exalted beings sometimes did.

On the ground level a tremendous golden image of Bel sat plumped before a golden table, its square stubby feet solidly planted on a bulging golden footstool. The reported poundage of gold in this artistic masterpiece passes belief. Outside the temple was more gold, molded into an altar, where only undefiled sucklings were cremated after having had their inno- cent throats cut. A more capacious altar, not of gold but merely gilded, accommodated the adult brutes. On this baser altar alone, in addition to all the bulls, cows, goats, sheep, and other livestock, 3,600,000 shekels' worth of frankincense were smoked up annually. How much this would be in tons may be left to historical experts. Whatever the answer turns out, no doubt it will be as impressive as Herodotus intended.

Some of what Herodotus describes of the citizens of Baby- lon has been partly verified from the surviving sculptures and monuments recovered by modern archaeologists from the mounds of crumbling rubble. The men depicted are clothed as Herodotus reports. Over a white linen tunic reaching to the ankles they wore a woolen garment and over this a white mantle. Their footgear was a rather clumsy kind of sandal. A verminous lot, apparently, the Babylonians neither cut their hair nor shaved. But they did curl their bushy beards. In compensation they massaged their entire but unwashed bodies with pungent oils, just as as the dainty belles of the eighteenth century did when they could get the deodorants. Each of the hairy gentlemen had his personal seal and sported a carved staff ornamented with his favorite device—"such as an apple, a rose, a lily, or an eagle." The apple, the rose and the lily recall the third of the Seven Wonders of the ancient world—the walls and hanging (tiered) gardens of Babylon. Judging by their perfumes and their lily staffs we might hastily infer that the Babylonians were rather on the effeminate side—a love of flowers has been popularly supposed to make men soft and gentle. Actually the Babylonians were among the bloodiest and cruelest warriors in all the bloody and cruel history of our race. In war they flouted even the merciful brutalities of the beasts. The Babylonians, being hardy and brutal, and being so tough—or at least those who survived—ignored personal hygiene and scorned physicians. Their substitute was socialized medicine in the most literal sense imaginable. When a man fell ill his friends carried him to a public square. Any passerby who had recovered from the man's sickness was obligated to comfort and advise the sufferer, and to share with him the mystery of his own recovery. When a patient died, as most did, he was preserved in honey.

As usual with primitive peoples and many of the ancients and some of the moderns, the Babylonians made much of sex. They were less finicky than the Victorians about it, or as some who have not yet heard of Freud still are. In one of his rare outbursts of enthusiasm, Herodotus declares that of all the established customs of the Babylonians, "those relating to marriage are the sanest." Once a year all the marriageable girls were rounded up and herded into a compound where the prospective husbands gathered to bid. There was something to suit every taste and every purse. Starting with the comeliest of all an auctioneer put up the girls for sale. When the prize beauty, Miss Babylon, had been sold the next fairest was offered, and so on till the deformed and the ugliest were knocked down to the highest bidders among the common people, who merely wanted a bedfellow and did not care how she looked in the morning. A purchase became the lawful wife of the buyer, who had to guarantee that he would respect her conjugal rights. The pay-off against the comeliest came after the sale of the ugliest. The poor culls were dowered with a split-up of the take on the sale of the beauties. Divorces were anticipated and provided for. If a couple could not get along together, the price of the wife was refunded and she returned to her parents. These "clean and happy customs" passed with the conquest of Babylon, when poverty was so dire that the poorer classes sold their daughters into prostitution.

The "foulest custom" of the Babylonians, according to Herodotus, sounds like a more or less refined survival of a primitive tribe. Once in her life every woman had to expose herself in the temple of the Goddess of Fertility and consort with the first man who tossed money in her lap. "Every woman" is meant literally. Married or single, a woman had to submit. The proud and the rich drove up to the temple in their luxurious chariots, but even if only a penny landed in her lap, the proudest or the richest had to acknowledge and discharge her religious obligation. Having accepted her man of the moment, the woman purified herself and returned home, never to be molested again. Herodotus observes that "the tall and fair" were quickly free to depart, while the dumpy and ugly sometimes had to wait three or four years. This much at least sounds authentic reporting.

To finish with the women, as Herodotus might say, I pass on to the miracle of the mule as he tells it, and his matter-of-fact recital of how the Babylonians of his day had acquired their great-grandmothers. When Darius, King of Persia, undertook the siege of Babylon the massive walls baffled him and he made but slow progress toward reducing the city. Confident that the walls could never be breached, the defenders pranced along the top, jeering at the besiegers: "Why don't you Persians go home instead of squatting there? You will take our city when mules bear offspring." In the twentieth year of the siege the miracle happened. One of the transport mules foaled. Or so the commanding general of the Babylonians testified. He had seen the offspring for himself. But possibly the general, being neither a veterinarian nor a geneticist, was unable to distinguish between a she-mule and a she-ass.[4] Whatever the cause, mule's foal or the wrath of Yahweh, Babylon fell. When, during the siege, rations began to run short, the defenders very sensibly strangled their women (but did not eat them), and disposed of the children—an ancient version of "women and children first." War was war in the good old times, and any man who left his wife or his children to the victors was either an imbecile or a cynic. After the fall Darius generously carted in a fresh supply of females–fifty thousand in all–pressed from the cleaner of the neighboring tribes, and these loving conscripts bred the Babylonians Herodotus knew.

The historic rules of Babylonia concern us only so far as their works may have demanded the development of some simple mathematics. Of all, Herodotus singles out two queens for special praise. The earlier of these, Semiramis—a melodious name loved by poets—is dismissed in one line in a schol-

---

[4]A geneticist referred me to an article by O. Lloyd-Jones in the *Journal of Heredity*, VII (1916), 494–502, "Mules that breed." This article includes a summary of some of the earlier literature on allegedly fertile mules. The geneticist is unconvinced; in fact he says he does not believe in miraculous mules. So probably it was Yahweh.

arly biographical dictionary as "the mythical founder of Babylon." Herodotus, who may have been romancing, says she built flood control dykes on the plain and finished some of the walls and temples of Babylon. Her successor by five generations, Nitocris, was the wiser woman according to Herodotus, and what he tells of her justifies his praises. (There was another Nitocris, "she of the rosy cheeks," Queen of Egypt.) Whoever and whatever this legendary queen of Babylon may have been, either she or some council of despots executed the great engineering works Herodotus attributes to her. Her case is somewhat like that of Shakespeare. If Shakespeare did not write his own plays, some other man of the same name did. In an account like this it does not greatly matter who did what. The important thing is that the work got done, and about that there seems to be no dispute. Traces of some of the vast engineering works attributed to Nitocris are still visible.

Nitocris (even if legendary) must have been a civil and hydraulic engineer of the first rank. Her masterpiece was the taming of the Euphrates to be her slave in peace and her ally in war. In accomplishing this she built many temporary bridges, dredged out an artificial lake for flood control and, as the work demanded, diverted the river out of its natural channel from time to time. On the more human side she had an acrid and quite womanly humor. Wishing her remains to lie undisturbed, she had this caution engraved above her tomb: "If any king of Babylon in time to come lacks money, let him open this tomb and take whatever money he desires. But let him not open it unless he lacks, or it will be the worse for him." Her tomb remained inviolate till Darius, who lacked no money, broke in. He found only the queen's shriveled body with this rebuke: "If you had not been insatiate of wealth and basely desirous of gain, you would not have violated the resting places of the dead."

33

In his account of the campaign of Cyrus against Nitocris' son, Herodotus mentions one detail which suggests that the ancients were less ignorant of scientific fact than is sometimes thought, whatever may have been their shortcomings, by our standards, in scientific theory and abstract mathematics. On his prolonged marches Cyrus took with him "very many four-wheeled wagons drawn by mules" bearing water in silver ewers. The water was from the river "Choapses which flows past Susa," and was boiled. How long did it take our own ancestors to find out that it pays on occasion to boil drinking water? Typhoid ("enteric") fever in the Boer War (ending 1902) cost more lives than bullets.

I must pass over the account of how Babylon the mighty, the supposedly impregnable, fell to a simple but ingenious trick that Nitocris herself might have imagined and may even have previsioned in her nightmares. Owing to the vast extent of the city, those in the inner section were unaware that they were captives till they "learnt the truth only too well while they were dancing and making merry at a festival." Unarmed, they were butchered.

Such, again as Herodotus would say, was Babylon. What the Hebrew prophets and St. John the Divine said about Babylon—"the Scarlet Woman," "the great whore"—sounds unduly harsh after Herodotus' temperate and sympathetic account. The Babylonians cannot have been quite as bad as their haters have made out. After all, the Babylonians have not yet been heard in rebuttal, and it is unlikely now that they ever will be. They impressed the Israelites of the "Captivity" but the Israelites did not impress them, and nothing has survived of what the Babylonians thought of their involuntary guests. Such in very brief were the Babylonians—a splended, human, brutal, and bloody people. What of their mathematics? In one word, it was great. Some of the evidence for this assertion will be given after I have disposed of a historical matter of considerable interest.

If the Babylonians were a bloody lot on occasion, the Assyrians surpassed them almost continuously in ruthless war and cold-blooded cruelty. It is impossible to find a human parallel for their unbridled ferocity; we have to go back to the carnivorous dinosaurs, long extinct, to match them.

The capital of Assyria was Nineveh, a name that survives as the epitome of all the crimes and violences associated with indiscriminate killings and mass murder. With one exception, Assurbanipal (Sardanapalus), the kings of Assyria were psychopathic terrors who killed, tortured, and oppressed for the mere hell of it.

Assurbanipal (669?–626? B.C.) had some of the attributes of a normal civilized human being. Being less warlike than his predecessors, he had time to cultivate the pursuits of peace. He is remembered for his library, which he assembled so that he might have "something to read." The manner in which he collected it is a preview of the gathering of the Alexandrian Library in the Museum—the foster mother of great mathematicians—by Philadelphus, the second of the Ptolemies to rule Egypt, in the third century B.C. (This is discussed in Chapter 5.) Assurbanipal ordered his agents to collect, by force if necessary, all the tablets they could find and bring them to Nineveh. There he established a corps of scholars and scribes to process the tablets and arrange them in some sort of intelligible order. Copies where needed were made. The library contained much—for instance, magic—of little intrinsic value; but the whole was a record of the historical and scientific knowledge of its time, including religion, philosophy, philology, medicine, astronomy, and mathematics. It has proved invaluable in reconstructing a picture of the whole civilization of Assyria and Babylonia. Assurbanipal had plenty to read.

The content of Babylonian mathematics is so extensive and so varied that I must limit my report to two items bearing on our main purpose of getting the theory of numbers down to Fermat's seventeenth century. The first item is "Pythagorean triples," to be explained shortly; the other, Diophantine equations of the second degree. The first is one of the finest things the Babylonians did in mathematics, and their best in the theory of numbers. We must leave aside their ingenious work in computation and other things irrelevant for the theory of numbers. However, not to abandon entirely those who may wish to know more about Babylonian mathematics (including geometry), I give references to three accounts, all clear and readable, and reproduce from Archibald[5] a skeleton summary of Babylonian geometry (and mensuration) of 2000 to 1600 B.C.:[6,7]

---

[5]R. C. Archibald, "Babylonian and Egyptian Mathematics," *American Mathematical Monthly (Outline of the History of Mathematics)*, Vol. 56, No. 1, Part II (January, 1949), pages 7–16.

[6]O. Neugebauer, *The Exact Sciences in Antiquity* (Princeton: Princeton University Press, 1952). With 14 plates. "Babylonian Mathematics, Egyptian Mathematics and Astronomy, Babylonian Astronomy," pp. 28–132. (It may be mentioned that Neugebauer, with his unique combination of first-hand knowledge of and extraordinary skill in the requisite dead languages involved, combined with an expert knowledge of the relevant mathematics and astronomy, has done more than any other man, living or dead, for the history of pre-Greek mathematics and astronomy.)

[7]O. Neugebauer and A. Sachs, with a chapter by A. Goetze, *Mathematical Cuneiform Texts* (New Haven, Conn.: American Oriental Society, 1945), XXIX. With 49 plates. (There are clear photographs of cuneiform tablets, also line transcriptions of tablets in cuneiform script.) The part we want here is in Chapter 3, especially pp. 40–41.

In the field of geometry the Babylonians of 2000 to 1600 B.C. used the following results in concrete cases, from which we have to infer that they were familiar with the general rules:

1. The area of a rectangle is the product of the lengths of two adjacent sides.
2. The area of a right triangle is equal to one-half the product of the lengths of the sides about the right angle.
3. The sides about corresponding angles of two similar right triangles are proportional.
4. The area of a trapezoid with one side perpendicular to the parallel sides is one-half the product of the length of this perpendicular and the sum of the lengths of the parallel sides.
5. The perpendicular from the vertex of an isosceles triangle on the base bisects the base. The area of the triangle is the product of the lengths of the altitude and half the base. Indeed the Babylonians may have thought of this result for the area of a triangle other than right or isosceles, since such a triangle may be regarded as made up of adjacent or overlapping right triangles; but there is no known example of this use of the formula.
6. The "Pythagorean" theorem; for example, for triangles with sides corresponding to the numbers 3, 4, 5; 5, 12, 13; 8, 15, 17; 20, 21, 29; and many more.
7. The angle in a semi-circle is a right angle.
8. The length of the diameter of a circle is one-third of its circumference ($\pi = 3$). The area of a circle is 1/12 of the square of its circumference (correct for $\pi = 3$).
9. The volume of a rectangular parallelepiped is the product of the lengths of its three dimensions, and the volume of a right prism with a trapezoidal base is equal to the area of the base multiplied by the altitude of the prism.
10. The volume of a right circular cylinder is the area of its base multiplied by its altitude.

11. The volume of the frustum of a cone, or of a square pyramid, is equal to its altitude multiplied by one-half the sum of the areas of its bases. It has been conjectured that the Babylonians had also the equivalent of an exact formula for the volume in the case of a square pyramid, namely

$$V = h\left[\left(\frac{a+b}{2}\right)^2 + \frac{1}{3}\left(\frac{a-b}{2}\right)^2\right],$$

where a and b are the lengths of the sides of the square bases. This was known to Heron of Alexandria 1700 years later, and reduces to the extraordinary formula apparently known to the Egyptians:

$$V = \frac{h}{3}(a^2 + b^2 + ab).$$

From his intimate and extensive knowledge of old Babylonian mathematics, Neugebauer in 1935 anticipated that at least the beginnings of a Babylonian theory of numbers, connecting the different parts of old Babylonian mathematics *by the investigation of the fundamental laws of the numbers themselves*, would be found. This anticipation was justified by a tablet which Neugebauer and Sachs[7] assign to the period between 1900 and 1600 B.C. It concerns "Pythagorean number triples," though it was written over a thousand years before Pythagoras was born. They should be renamed Babylonian number triples. But as the misnomer is so deeply embedded in the mathematical vocabulary that it would take too much time and effort to dig it out and discard it, I shall adhere to it.

If l, b denote the longer, the shorter, legs of a right-angled triangle, and d is its hypotenuse (longest side), then

$$l^2 + b^2 = d^2,$$

and (l, b, d), for l, b, d integers, is a *Pythagorean triple*. If l, b, d are *coprime* (have no common divisor greater than 1), the triple (l, b, d) is now called *primitive*. If p, q are coprime integers and not both odd, and if p is greater than q, *all primitive triples are given, without duplications by*

$$(2pq, p^2 - q^2, p^2 + q^2).$$

For example, take $p = 2$, $q = 1$, and get $(4, 3, 5)$, $4^2 + 3^2 = 5^2$, or $16 + 9 = 25$. A number is called *regular* if it is of the form $2^r 3^s 5^t$, $2, 3, 5$ being the prime divisors of the sexagesimal base 60. Denote this number by $[r, s, t]$. The reciprocal of a regular number in the sexagesimal system is a terminating number. The makers of the table of triples evidently wanted terminating numbers, as they used division (multiplication by reciprocals) in their calculations. For example, $p = [1, 3, 0]$, $q = [0, 0, 2]$, or $p = 54$, $q = 25$, so that the triple $(l, b, d) = (2700, 2291, 3541)$.

That the Babylonian mathematicians of the second millennium B.C. should have had a method for producing results like these on Pythagorean triples is as astonishing as the ingenuity which deciphered them from the crabbed script of a long-dead language. Until this tablet was read, there was no evidence that a rule was known for finding Pythagorean triples similar to that proved by Euclid (third century B.C.) in Lemma 1 to Book 10 of his *Elements*.

The other item relevant to the theory of numbers concerns indeterminate equations of the second decree to be solved in integers. These belong to Diophantine analysis, after Diophantus of Alexandria. The European mathematicians of the seventeenth century and their successors down to our own times attacked, but did not always solve, many special equations of this kind of other degrees. There is no general, comprehensive theory.[8] The specimen exhibited presently is Babylonian, of about 2000 B.C. It is of interest because it takes a step beyond the Pythagorean $a^2 + b^2 = c^2$, by affixing a coefficient, here 2, to one of the squares. Such apparently slight modifications of a solved equation may result in an equation which nobody could solve, because it is impossible. For example, the change of 2 to 3 in $a^2 + b^2 = c^2$ yields the unsolvable $a^3 + b^3 = c^3$ —one of Fermat's.

The equation[9] is equivalent to

$$x^2 + y^2 = 2z^2,$$

to be solved in integers. For the practical problem leading to this equation, I must refer to Dr. Gandz' paper, as the statement is somewhat long and complicated. It is always interesting when a nontrivial equation (like this one) can be solved completely. With a little ingenuity it can be shown that the complete integer solution is

$$x = k(t^2 + 2tu - u^2),$$
$$y = k(t^2 - 2tu - u^2),$$
$$z = k(t^2 + u^2),$$

where all the letters denote integers, k is any integer, t, u are of opposite parities (one even, the other odd), and are coprime. A Babylonian, or even Diophantus, might have tried $x = \sqrt{2z^2 - y^2}$ directly from the equation. Then he would have seen that $2z^2 - y^2$ must be a square, say t^2, so $2z^2 - y^2 = t^2$, or

---

[8]It may be impossible to construct one.

[9]From the paper by S. Gandz, *Osiris*, VIII (1948), 13–40. There are others in the same paper.

$t^2 + y^2 = 2z^2$, of the same form as the original equation, and the unlucky man would be back where he started. If the affixed coefficient above were not 2 but 11 or 23, we should be in real difficulties.

2. Egypt

In discussing the Egyptians, I shall consider first a few features of their mathematics, and then some social details of interest. The order here is thus the reverse of that taken for the Babylonians. The following preliminary remarks may clarify the account and forewarn the reader what not to expect.

On crossing into Egypt, we (at any rate, I) run into several roadblocks that cannot be crashed through, but must be obviated or somehow circumvented. In such an impasse we may go either to the right or the left in getting past a block. When we have passed it, there are usually two ways of proceeding. The roads diverge and lead to different places. As there may be no means of bringing the two together, the only way of getting anywhere seems to be to describe both places, and let everyone, from such knowledge as he has, choose which of the two seems preferable to him. Or, perhaps more sensibly, he may ignore them both and proceed to something less dubious. For example, two accounts of the social climate of Egypt during and immediately after the building of the Great Pyramid are mutually contradictory. I give both. The older, traditional one, follows Herodotus, and presents Cheops (or Khufu), the builder of the Pyramid, as a brutal and callous tyrant. A much more recent one pictures him as a benevolent despot, as kindly as an old gentleman in his dotage. Possibly a Hegelian synthesis might reconcile the two accounts, but I shall not attempt it. There is a similar, but milder, disagreement concerning an alleged decline of Egyptian culture in the long span from the pyramid age to the end of Egypt as an independent state in Roman times.

Another obstacle blocks our way, more formidable than most purely historical difficulties—although this one too might be classed as historical. The word *mathematics* has occurred several times in the narrative so far, without any indication of what it means. "The history of mathematics, as of any science, is to some extent the story of the continual replacement of one set of misconceptions by another."[10] This remark is particularly pertinent when we attempt to answer the question "What is mathematics?" All the difficulty is concentrated in that too prolific copula *is*—the progenitor of philosophies from Plato to the present. For the moment it suffices to remark that mathematics today is not what it was the day before yesterday, whatever it may "be." Axiomatics worked once for a majority. It is still acceptable to many.

For an extended discussion of the implied difficulties in this brief notice, see R. L. Wilder, *Introduction to the Foundations of Mathematics* (New York: Wiley and Sons, 1952). The classical axiomatic method described here is sufficient for Greek mathematics and for much of modern mathematics till 1912. But no mention of intuitionism (1912–   ) and metamathematics has been made. Greek mathematics was largely visual and intuitive in the loose philosophical sense of the Greeks. It was good enough for their geometry. Intuitionism in the current technical sense is something quite different and offers a coherent account of *numbers*, which the Greeks never gave.

---

[10] F. De Sua, "Consistency and Completeness, a résumé," *American Mathematical Monthly*, LXXIII (1956), 295–305.

Even on their own axiomatic level the Greeks, for example Euclid, as will appear later, were on the same empirical, visual level as the Egyptians. The greatest Greek mathematician of them all, Archimedes, never proved anything by any civilized current standard. His "mathematics" was purely Egyptian protomathematics. This verdict is not mine; it is a commonplace to men who are mathematicians by trade. So let us temporarily accept the stale myth of the (intellectual) "glory that was Greece" and get on with something less glorious. What has been said may be retained as a reminder that mathematics is alive.

Mathematics since the time of the Pythagoreans (sixth and fifth centuries B.C.) has meant, for most mathematicians, if not for others, the strict logical deduction of mutually consistent statements ("propositions") from explicitly stated assumptions ("postulates," "axioms") accepted without argument as "true." This is the method of "axiomatics," and should be familiar from its application to the observed, empirical properties of plane figures (triangles, circles, etc.) in the first school course in geometry. In short, it is the method of deductive proof. There is no hint of proof in either Babylonian or Egyptian mathematics as so far discovered.

If what *mathematics* means to many mathematicians is accepted, what were the Babylonians and the Egyptians doing? Whatever it may be called, it was not mathematics. Nor, for that matter, is the algebra and geometry of the usual first school course. Yet all that the ancients before Greece did, and the school children continue to do, is commonly called mathematics and, it seems to me, justifiably, if we qualify *mathematics* by two adjectives, *pure*, *applied*, and know which we are talking about. Pure mathematics is accepted by a majority as above described. Applied mathematics is an empirical science which appeals to sensory experience for its validation. Some schoolbooks of fifty or more years ago asserted that geometry is a science on a level with the physical sciences. The statements of applied mathematics supply the data for pure mathematics to take hold of and reduce to strict logical coherence.

The passage from the applied to the pure is sometimes long and difficult. To take an example, it is stated (in effect) by Euclid and allegedly proved in the schoolbooks that the area of a plane triangle with base b and altitude h is obtained by taking half the product bh (in the appropriate units of length), or area $= \frac{1}{2}bh$. This apparently was known to the Egyptians. How is it to be proved? Clearly we must have some agreement concerning the meaning of *area*. This was long in coming. Granting that we have got thus far, we may demand that the proof can be carried through in a *finite number* of steps, but an *unending sequence* of steps may turn out to be required before a proof is obtained. The second of these is far beyond a school course, and indeed beyond any but fairly advanced college courses. The first is therefore desirable, if $\frac{1}{2}bh$ is to be lifted from applied to pure mathematics. Such a proof was first constructed only in this century, by D. Hilbert; so we need not be too contemptuous of the Egyptians with their rule of thumb —which worked well enough for their purposes.[11]

I have dwelt on $\frac{1}{2}bh$ because a similar but much more difficult question occurs in the most extraordinary thing the Egyptians did in geometry—the correct formula for the volume of a truncated regular pyramid on a square base, which has

---

[11]For the proof see D. Hilbert, *Grundlagen der Geometrie* (7th ed., 1930), Chap. 4, pp. 67–78. There is an English abstract from earlier editions of the *Grundlagen* in T. L. Heath, *The Thirteen Books of Euclid's Elements* (Cambridge: Cambridge University Press, 1908; 3 vols.) I, 328

already been cited (No. 11) in Archibald's list. This one, as will appear, cannot, like the area of a triangle, be lifted by *anybody* in a *finite* number of operations from the applied to the pure.

Other notable achievements of the Egyptians in applied mathematics are their calendar—"the only intelligent calendar which ever existed in human history" (Neugebauer,[6] pp. 80–81)—and their measurement of time.[12] I shall merely state the final outcome of what must have been centuries of observation and experience preceding this unique "intelligent calendar": "12 months of 30 days with an additional 5 days at the end of each year." The thirty-day month was divided into three periods of ten days each. To impose this modicum of sense on our own civilization, the calendar reformers will have to upset all the governments and religious organizations in the world.

The heavens, to the Egyptians, were a perpetual year-clock, providing a constant measure for the length of the year. But the lengths of the days and nights vary with the seasons. By what means could the Egyptians judge how much time had elapsed between sunset and sunrise? This interval, like the length of the day (sunrise to sunset), was divided into equal parts, "hours." To measure these, they invented (before 2000 B.C.) the water clock (clepsydra), which measured equal intervals of time with sufficient accuracy. The volume of water flowing into or out of a container is proportional to the weight of the water, which can be easily determined. A scale indicated how much had flowed out at a hole at the bottom of the

---

[12] For another account of the calendar, see R. A. Parker, *Journal of Calendar Reform*, XXV, June, 1955.

45

container. Obviously they had what was required as a measurer of time, day or night. In our own civilization with its mechanical clocks, including pendulum clocks, an elaborate mathematical science (horology) has developed since Galileo in the sixteenth century measured the time of oscillation of a lamp in a church against the beatings of his pulse. The Egyptians used only the simplest arithmetic and great ingenuity.

How does the applied mathematics of the Egyptians compare with our own? Methodologically, there seems little, if anything, to choose between them. In only a few isolated and comparatively unimportant provinces of the vast territory covered by mathematical physics, has even an attempt been made to supply a system of postulates. There is no reason why the pure, axiomatic technique should be adopted by the applied mathematicians; they continue as they have done for centuries to get on well enough without it, even as the Egyptians in their day did. So we may temper any scorn we may have for them with a little common sense, and set our own house in order—if and when a clean-up is indicated. One such occurred in 1905, when Einstein added a postulate to Newtonian mechanics, thus precipitating the atomic age; another in 1900, when Planck quantized energy, making possible the quantum theory. These however can hardly be attributed to pure mathematics as defined here; they were insights into the observable structure of the physical universe.

The first item of more than passing interest in Egyptian mathematics is their peculiar way of handling common fractions. It is theirs; no other people who did likewise is known, till the Greeks appropriated it. In itself it is of little interest for mathematics; but two problems it suggested to workers in the theory of numbers 3600 years after the Egyptians manipulated their fractions, are of decided interest. The first problem, as we shall see, was generalized and solved in 1880; the second has been solved only in special cases.

A *unit fraction* is a fraction of the form 1 divided by an integer, such as $\frac{1}{28}$, $\frac{1}{776}$. A table in the Papyrus Rhind, of about 1680 B.C., gives the sum equivalents in unit fractions of 2 divided by each of the odd numbers from 5 to 101; $\frac{2}{3}$ is not resolved into unit fractions, but is denoted by a special symbol —the Egyptians had no notation for a fraction with a numerator exceeding 1. As an example,

$$\frac{2}{97} = \frac{1}{56} + \frac{1}{679} + \frac{1}{776}$$

is a decomposition of $\frac{2}{97}$ into a sum of unit fractions. If the problem is to decompose $2/n$, n being an odd integer, into a sum of unit fractions, it can be done in an infinity of ways, and there is nothing to suggest why out of this unordered swarm the Egyptians chose a particular one, when almost any other would have done as well. From a modern point of view, this is most unsatisfactory. Further, the restriction to 2 of the numerator of the fraction to be decomposed is a blemish that should be removed, if possible—but the Egyptians could not, because a fraction with a numerator exceeding 2 was not in their arithmetic. Sylvester in 1880 removed the blemish and gave a simple, nontentative algorithm like that of continued fractions in older school algebras, for producing a *unique decomposition of any proper fraction (fraction less than one) into a sum of unit fractions.*[13]

---

[13]J. J. Sylvester, "On a point in the theory of vulgar fractions," *American Journal of Mathematics*, III (1880), 332–335, 388–389; reproduced in J. J. S., *Mathematical Papers*, III (1909), 440–445.

Sylvester calls the reciprocal of an integer a "simple" fraction (instead of a "unit" fraction); "any other fraction, whether rational or irrational, may be termed complex, it is to be understood that only...fractions greater than zero and less than unity will be considered." He proves that, and gives a straightforward, usable method for producing the result: "every complex rational fraction [in the sense he has defined] can be expanded, and only in one way, under the form of a finite series." As examples,

$$\frac{4699}{7320} = \frac{1}{2} + \frac{1}{8} + \frac{1}{60} + \frac{1}{3660}; \quad \frac{335}{336} = \frac{1}{2} + \frac{1}{3} + \frac{1}{7} + \frac{1}{48}.$$

Sylvester tells how he became interested in the problem:

The preceding matter was suggested to me by the chapter in Cantor's *Geschichte der Mathematik* which gives an account of the singular method in use among the ancient Egyptians for working with fractions. It was their curious custom to resolve every fraction [of the form $2/n$, n odd] into a sum of simple fractions according to a certain traditional method, not leading, I need hardly say, except in a few of the simplest cases, to the expansion under the special form to which I have, in what precedes, given the name of a fractional *sorites*.[14]

The second problem suggested by the "curious custom" of the Egyptians is to solve completely in positive integers, where m, n are any given positive integers and $x, y, z, \ldots$ are to be

---

[14] Because Sylvester's algorithm resembles the process so named in classical logic.

found, indeterminate or Diophantine equations of the types,

$$\frac{m}{n} = \frac{1}{x} + \frac{1}{y}, \qquad \text{or} \quad mxy = n(y+x);$$

$$\frac{m}{n} = \frac{1}{x} + \frac{1}{y} + \frac{1}{z}, \qquad \text{or} \quad mxyz = n(yz + xz + xy),$$

and so on for any number of unknowns $x, y, z, \dots$. Many special cases have been disposed of in the past two centuries, but the general case is still outstanding.

Though the Egyptians' miscellaneous contributions to mathematics hardly touch the theory of numbers, a very few items may be mentioned to give some indication of their general mathematical level. For fuller accounts see Archibald[5] and Neugebauer.[6] The Egyptians gave correctly the formula for the volume of a cylindrical granary. Simple problems led to what we would set up as equations of the first degree; for example, $\frac{2}{3}x + \frac{1}{2}x + \frac{1}{7}x = 33$. As they had no algebra, they must have got the answer a harder way.

Their arithmetic was wholly additive; they had no multiplication. But they got around this deficiency by expressing one or two numbers to be multiplied in the binary scale of notation (by sums of powers of 2). For example, $6 = 2 + 4 = 2 + 2^2$; $9 = 1 + 8 = 1 + 2^3$. The decomposition of a number into a sum of powers of 2 is unique (1 is the zeroth power). To compute 11×7, say, express 11 in the binary scale as $11 = 1 + 2 + 8$, and make the table

✔	1	7
✔	2	14
	4	28
✔	8	56
	16	112

Take $1, 2, 8$ from the left, and their opposite numbers 1×7, 2×7, 8×7 from the right, obtained by successive doublings; add the latter:

$$7 + 14 + 56 = 77 = 11 \times 7.$$

Binary arithmetic since the 1930s has contributed notably to the investigation of perfect numbers, Mersenne numbers, and Fermat numbers (discussed in Chapters 10, 11). This has come about through the invention of modern computing machines, in particular the digital computers, whose electronic mechanisms indicate the "yes-no" type of answer given by the two-valued classical (Aristotelian) logic. What these machines can do in a matter of hours for the numbers mentioned, also for exploring Fermat's Last Theorem, would take (in some instances) centuries with a desk computer. The Egyptian use of binary arithmetic was an inspiration for its time. It is a "natural" tool for investigating certain intrinsic properties of numbers. For the same purpose the decimal notation is an unnatural artificiality. What has the number 10 to do with the innate properties of the integers?

English school texts of as late as the nineteenth century had many problems on "alligation." An Egyptian application, hardly suitable for the Victorian age, asked for the proportions of different brews of beer to be mixed to produce a beer of given potency; or, given the amounts mixed, to determine the strength of the mixture. In passing, I note that the Egyptians appreciated the peculiar virtues of alcoholic beverages. Drunkenness was not a disgrace. In a picture of a party in the eighteenth dynasty, a maid, handling round cups of wine, says persuasively to a lady guest, "Drink this and get drunk." The lady replies with great animation, "I shall love to be drunk."[15]

With that invitation we shall leave the uninhibited Egyptians to the enjoyment of their rather meager elementary mathematics, and lay the historical ghost of the "harpedonaptae," or "rope-stretchers." The nearest current equivalents of these mysterious beings are the chain men in a surveyor's gang.

Geometry etymologically is from the Greek for "earth-measurement." The land surveyors are depicted on the monuments going about their business of stretching cords to measure distances. They were the "rope-stretchers" whose precise function for long puzzled historians of mathematics. (Had they consulted the Book of Job they would have guessed.) One sketch depicts a surveyor accompanied by a lady assistant, possibly a queen or goddess, engaged in the abstruse mystery. It was once thought that the rope-stretchers used a marked cord $3 + 4 + 5$ units in length to lay down a right-angled triangle with sides $3, 4, 5$, of obvious utility in orienting buildings such as temples and pyramids under construction. If this were so, the Egyptians would share with the Babylonians the credit of having taken the first long stride toward a theory of numbers as distinguished from mere practical calculation. But it appears that the Egyptians were unaware of the arithmetical connection of 3, 4, 5 ($3^2 + 4^2 = 5^2$) with the sides of a right triangle. They also were unacquainted with even a single instance of the equation $a^2 + b^2 = c^2$ in integers a, b, c, and moreover they had no knowledge of the equation's geometrical

---

[15]From Margaret A. Murray, *The Splendour That Was Egypt* (London: Sidgwick and Jackson Ltd., 1949), p. 119. The divinity of the gods was attributed to inebriation.

significance (the Pythagorean theorem). It seems rather a pity that the old legend of the rope-stretchers laying down right angles has been exploded by modern scholarship. But we have this consolation: even the highest-powered modern scholarship is unlikely to demolish the Great Pyramid. It will take a super-bomb to do that, and there may be no tourists to view the ruin.

Some historians of mathematics have insisted that the mathematics of a people is a reliable index of the culture. This may be true for the Babylonians; it is contradicted by the Egyptians and the Chinese. Both of these long-lived peoples have only a comparatively poverty-stricken mathematics to show beside their magnificent achievements in art. It is merely a facile guess, but it would seen that the Egyptians fell so far short of the Babylonians because of a simple lack of curiosity and impractical imagination. Consequently, they were blighted in their mathematics by the curse of a narrow practicality which they never outgrew. What could not be immediately translated into stone temples, pyramids, irrigation ditches, bookkeeping, crops, and religion was of but slight interest to them. A primitive attempt at engineering, taxes, food, trade, ethics, and a blind reverence for the oppressive past seem to have prescribed the boundaries of their intellectual activity. The similarity here to the post-World War II attitude toward basic science is obvious and needs no elaboration. Babylon warred itself into suicide; Egypt lapsed into a more or less peaceful decline. (The other side of this picture will appear later.) When Herodotus visited Egypt it was little more animated than a city of death. The dead were everywhere, from the tombs of the kings to the thoughts of the lowliest worker hemmed in and stultified by innumerable traditional tabus and religious observances whose significance had long been forgotten. This, that, or the other had to be done according to its rigid formula because men and women who had been mummies for two thousand years or more had done it so. The spiritless people dozed away their lives in the shadow of an antiquity which had long since lost all meaning for the living. Fertility feasts and religious pantomimes supplemented by enough to eat and plenty to drink—the Egyptian version of "bread and circuses" —kept the people apathetic and docile. They lost whatever mathematical curiosity they may once have had. After a promising start in the second or late third millennium B.C., Egyptian mathematics remained stationary for 2000 years. It never really came to life again after its premature mummification. The native Egyptians had no part in the mathematics of the Alexandrian school of the third and second centuries B.C. A possible explanation for this premature fossilization appears in what Herodotus reports of the Egypt of his day. "Nowhere," Herodotus says of Egypt, "are there so many marvelous things, nor in the whole of the rest of the world are so many works of unspeakable greatness." He was referring to the pyramids, the temples, the causeway of polished stone, the sculptured tombs hacked into the living rock, the avenue of sphinxes, the colossal granite statues polished to the smoothness of glass, the huge obelisks, the incredible engineering works, and in fact all the marvels whose time-worn remains can be seen today. With only a rudimentary science and the simplest mathematics, how were all these "marvelous things" possible? The answer may be, as Herodotus implies, the ruthless determination of egomaniacal and death-devoted tyrants to perpetuate themselves and to reach out for a personal immortality at the expense of an entire nation reduced for generations to grinding slavery that exhausted the people and degenerated their stock. The slave-driving tyrants themselves were enslaved to a cult of death. The opposite side of this depressing account asserts that professional Egyptologists see no evidence of a universal decline in culture. The example of a stationary mathematics indicates nothing for the general level of Egyptian civilization. Change was continuous in many different departments, for example writing. Life was still active and creative, except in mathematics. The disagreement between the two appraisals may possibly be credited to the priests who misled Herodotus for purposes of their own.

We must pass on to the pyramids. The pyramid age lasted from 4750 to 3000 B.C. (by one chronology), after which no more were built. There were several of them, many mere nubbins compared to the giant of them all, the Great Pyramid of Cheops, built about 3050 (2900?) B.C. Some figures will give an idea of this bulkiest of all memorials to the vanity of a human being. The dimensions have shrunk since the builders finished their task and abandoned their masterpiece to the mercies of time, weather, and robbers. The base originally covered between twelve and thirteen acres; now 750 feet square, it was 768 feet to the side before the peasants began quarrying in the huge mass, 3,057,000 cubic yards of masonry, for stone they could use. The original height was 482 feet, now it is 451. The apex vanished long ago, the top is now a flat area about 12 yards square. In fact, the whole now is a truncated pyramid on a square base. The original outer casing of polished white limestone followed the apex, Osiris and Cheops only know where. Today the perfection that was the Great Pyramid in its prime presents a somewhat untidy and dilapidated spectacle of 203 steps, up which avaricious guides lug winded tourists. If Cheops is aware of what has happened to his supposedly indestructible monument, he must be turning cartwheels in his unknown grave. His sarcophagus was looted soon after it received him. But possibly, as we shall see, he escaped extinction after all in his "solar boat" which was to have transported his soul, at least, underground to the Judgment Hall of the Dead. The capacious boat, in a remarkable state of preservation, was discovered only in 1954. Archaeologists and others had been tramping over the hiding place for years without suspecting what lay under their feet. Cheops evidently missed the boat; no vestige of him, body or soul, was discovered. Can it be that the judges of the dead weighed his heart—according to their custom—and found it wanting? It is not impossible.

Passing to geometry, I shall consider the Egyptian formula for the volume of a truncated regular pyramid on a square base, the high-water mark of their mathematics, and in doing so abstain from adding one more guess to the several extant concerning what *may* have suggested to them the problem of finding the volume, *not* its solution.[16] However the Egyptians got the correct result, it was an empirical discovery of a very high order. A rigorous demonstration, as proved in the present century, *necessarily* demands the use of the integral calculus, or some equivalent limiting (unending, infinite) process.[17]

---

[16] For hypotheses on this, see J. L. Coolidge, *A History of Geometrical Methods* (Oxford: Oxford University Press, 1940), pp. 8–10.

[17] D. Hilbert, *Deuxième Congrès International de Mathématiciens*, 1900 (published 1902), *Problèmes Futurs de Mathématiques*: twenty-three problems, of which the third (pp. 74–75) is to prove, without appeal to infinite processes, such as the Greek (Eudoxian) method of exhaustion, or the use of continuity, that two tetrahedra (triangular pyramids) of equal bases and equal altitudes have equal volumes. The difficulty is to produce a proof in a *finite* number of operations. We noted that Hilbert proved the corresponding result for plane triangles, using only a finite number of operations. Gauss in 1844, *Werke*, VIII, 242, was aware of the difficulty in the case of solids. Hilbert later said he believed the problem impossible.

M. Dehn, in 1902, proved the impossibility in his paper "Über der Rauminhalt," *Mathematische Annalen* (1902), Bd. 55.

Why, it may be asked, if a proof by appeal to infinite processes is possible, should anyone have tried to find a finite proof? For the sufficient reason that the use of the infinite (as in continuity) introduces new and unresolved, possibly unresolvable, difficulties. There is not space here to go into this.

In some of the schoolbooks on elementary solid geometry, there is a supposed proof of the Egyptian formula by cutting the frustum into three tetrahedra, which can be displayed so as to exhibit equal bases and equal altitudes. But this refers the problem back to the one which is impossible by finite means.

The formula is just the kind of mathematics the pyramid-building Egyptians might be expected to have been interested in. Suppose for example that a pyramid was built up to only half its final height. The king whose mummy was to be sealed up in the secret chamber of the completed pyramid to wait the judgment of the dead would naturally wish to know how long it would take to finish his tomb. If he knew the volume of a whole pyramid he had only to subtract the volume of the truncated part already completed to get the volume of what remained to be built. But did he know the volume of a whole pyramid? To us it is an immediate deduction to get the volume of a whole pyramid from the formula for a truncated pyramid. So far there is no evidence that the Egyptians ever took this short step to the solution of a problem which must have been of importance to them. It was natural for them to miss this step, as they could have had no conception of a zero area. Lacking this final detail, the king may never have satisfied himself that his tomb would be ready for him when he was ready for it. But even if the builders beat him by only an hour, it may have been some comfort to him to keep a tally of how much stone was going into the pyramid as the forbidding mass rose higher and higher from month to month, and for this the truncated formula was sufficient. There is no evidence that any of the pyramid builders were familiar with the formula. But if none were, it seems strange that the narrowly practical Egyptians should have troubled their unmathematical heads over a really difficult mathematical problem.

Whether the pyramids influenced Egyptian mathematics may be doubtful, but that these colossal masses cast their oppressive shadows on the lives of the Egyptian people for generations has been claimed. According to Herodotus, Cheops, the builder of the Great Pyramid (flourished about 2900 B.C.), reduced the people to "utter misery." Until his time Egypt was well governed and prosperous. To ensure a constant supply of slave labor, Cheops closed all the temples and enslaved the entire population in the building of his supposedly eternal tomb and everlasting monument. Swarms of slaves, teamed up like ants, dragged the huge blocks of stone required for the pyramid from quarries in "the mountains of Arabia." The blocks were then ferried across the Nile on specially constructed rafts. They were next dragged up on rollers by crude brute force to "the mountains called Libyan." The slaves worked in rotating gangs of a hundred thousand men, each for spells of three months.

For ten years the people were "afflicted" with the building of the causeway from the Nile to the site of the pyramid, over which the stone blocks, thirty or more feet long, intended for the main structure, were dragged to their final destination. According to Herodotus, this task was but little less hellish than the building of the Great Pyramid itself. The causeway was "five furlongs long, ten fathoms broad, and five fathoms high," all of polished stone. The greatest sculptors and the most highly skilled stone-workers toiled at carving and polishing statues to perpetuate their overlord's taste in art. These masterpieces were disposed at pleasing intervals along the

57

smooth road of sweat and lashes. The Great Pyramid, dating from about 2900 B.C., took only twenty years to build. This seems incredible, but it is in the accepted record.

Modern engineers have wondered how the huge blocks were lifted and eased into place. Herodotus does not say, but it appears from modern research that as the work progressed long ramps of sand sloping up to the faces of the pyramid as far as they were constructed at any stage were heaped up, basketfuls at a time, and up these the slaves dragged the massive blocks on rollers for the next tier. When the top stage was reached the pyramid was buried under a succession of longer and longer ramps. When the final labor, beginning at the apex, was completed, the thousands and thousands of tons of sand burying the pyramid were carried away in baskets and spread out evenly over the surrounding countryside.

The first stages of the construction were by massive steps. The slaves then eased in triangular prisms of polished limestone to make the four surfaces plane and dazzlingly white. The topmost part was smoothly cased first, then the next below it, and so on to the lowest course. Before the ordinary Egyptians, after King Cheops was entombed, carted away the smooth outer casing of limestone to build their own lime burners and to fence their cattle pens and pigsties, the Great Pyramid was as uniform and symmetrical as if it had been sliced out of solid stone by a gigantic knife. The surfaces were so smooth that an ant might have difficulty in climbing to the top without a slip. As for Cheops, he seems to have been unaware that thieves break in and steal, and that no safe or sarcophagus devised by human ingenuity is beyond human ingenuity to crack and loot.

Herodotus evidently disliked Cheops. This great king, he says, was so wanting in common human decency that he dedicated his own daughter to prostitution to raise money for his tomb. This at any rate is the story as told to Herodotus by his priestly tourist-guides, the dragomans of his day. The daughter was as practical as her father; she exacted a tip of one block of stone for each of her favors. The middle pyramid (so the priests told Herodotus) was built from the stones thus earned. Each side of this pyramid measured a hundred and fifty feet. If the priests were not romancing, Miss Cheops must have been quite a busy young woman.

Slavery with its attendant ignorance and misery continued into the reign of Chephren, Cheops' brother, lasting in all a little over a century. All this time the temples remained closed. The tyrants responsible for this inhuman oppression earned the hatred and contempt of the people for generations, and not very long after the kings had themselves sealed up in their supposedly thiefproof tombs some of their sarcophagi were violated and plundered, and whatever remained of them was thrown out on the sands as so much worthless rubbish. Others of their kind have been less fortunate. Their shriveled mummies bore idle tourists wandering through the museums of our Western culture, looking for a bench to sit down on and rest their aching feet. "Thus passes the glory of the world"—but not the formula for the volume of a truncated square pyramid.

Not all of the rulers were as harsh and shortsighted as Cheops and Chephren, otherwise there would soon have been no slaves to oppress. The Nile was the ultimate despot of Egypt. Its annual floods buried fertile fields under feet of silt and left marshes that had to be drained. Herodotus, incidentally, speculates on how long it would take for the river to silt up completely, and estimated between ten and twenty thousand years. Canals for drainage and irrigation converted what had been "a horse-and-cart country" into one in which the only sensible transportation was by waterways. When the annual floods receded, towns not close to the Nile had only brackish water from wells, so canals from the river were dug to supply the inhabitants with fresh water. All this prodigious labor required endless bookkeeping of a rather rudimentary kind. Some of the overseers kept accounts as detailed as those today on a major engineering job. Thus Herodotus reports that "there are records in the Great Pyramid showing how much was spent on purges, onions and garlic for the workmen." Of such are the origins of arithmetic.

When the kings became practical in the art of rule and began to let the people breathe freely, a parasitic swarm of priests and holy men took over and sucked a juicy living from the populace in payment for allegedly placating the numerous gods. The priests were not charged solely with keeping the gods pleased. More importantly they conserved and transmitted whatever mathematical and scientific lore the Egyptians had. Realizing that knowledge is power, the priests clutched all this subversive lore tightly in their own fists. They were the Atomic Energy Commission of their time. Though not actually money to them, knowledge was the equivalent of money in fat living and scholarly leisure. In such matters as surveys to readjust the boundaries of fields blanketed under silt by the annual inundations of the Nile, it might have been inexpedient to let the peasantry suspect too much of what was going on. But, according to Herodotus, there seems to have been very little graft, and probably the mysteries of measuring a plot of ground were as incomprehensible to the serfs of Egypt four thousand years ago as the intricacies of nuclear physics or the income tax are to the voters of the U.S.A. since the atom bomb. Docile ignorance and a moderated oppression kept the Egyptian peasant of four thousand years ago loyal. The horde of smooth-shaven and plump priests and indifferent savants, who did no work as the peasants understood work, does, however, seem to have been more numerous than was necessary, and the diggers of ditches and cultivators of fields carried this mob of holy men and erudite idlers on their bowed backs. There were tens of thousands of these superior and supercilious sycophants. Taxes in kind—the fattest geese, the choicest cattle, the flower of the cereal crops, the freshest fish and the best of the wines and beers—were pressed on the sanctified racketeers in grateful tribute for celestial services promised, but not rendered, in this life. It is scarcely surprising that under such a regime Egyptian mathematics and science fossilized nearly two thousand years before Egypt became a decadent appendage to the ramshackle empire of Alexander the Great. The virile Macedonians, not the effete Egyptians, made Alexandria the mathematical capital of the ancient world in the centuries from Euclid to Diophantus.

The account given here of slavery and oppression under Cheops and his successors substantially follows Herodotus. But since his day, much that he did not know has been discovered. It will suffice to mention one most significant detail. The temples were closed during the building of the Great Pyramid allegedly *to give the tillers of the soil work and food during the season of the Nile inundation when agriculture was impossible*. When the waters subsided, the people went back to productive labor, planting the seed (they trod it into the mud) for the next necessary crop. Before long, the new crop was ready for harvesting, and prosperity boomed. So Cheops, by giving the people work, is acquitted of mere wilful tyranny. He may have caught his solar boat after all. If Herodotus were right he might have missed it.

It seems curious that the Babylonians and the Egyptians, contemporaries for many centuries and not very distant neighbors, should have differed as widely as they did in their customs. Passing over some rather indelicate details, especially about the women, I shall report a few of the differences

61

between the two peoples as Herodotus observed them, and I shall take the women first.

O, true image of the ways of Egypt that they [the young men] show in their spirit in their life! For there the men sit weaving in the house, but the wives go forth to win the daily bread.[18]

In Babylon the men followed what we usually consider manly pursuits from trade to war. In Egypt the women were the buyers and sellers while the men sat at home weaving. This is hard to believe, but so it is reported, and on the monuments there are portrayals of men at their looms. After all, if male couturiers design our women's clothes today, why should not the men of Egypt have woven the stuff of clothes without thereby compromising their masculinity? Herodotus notes what he thinks an outlandish detail in the weaving. The Egyptians, he says, were unique in pressing the woof downwards instead of upwards. Though the women did not weave, not all were exempted from hard work.

The men indeed were little more than domestic servants and household pets. Their women, of the higher classes, painted them up like dolls and blackened their eyelashes with kohl. These superior women ran the businesses, banking, law, agricultural administration, and international affairs, while the men cooked, reared the children and wove their clothes. The women of the lower classes did not have it so soft as their aristocratic sisters. These depressed drudges tilled the fields, tended the livestock, hauled water from the Nile, did all the janitor work in the temples, and saw that the sacred cats were

---

[18]Sophocles, *Oedipus at Colonus*. The translator (Jebb) says Sophocles probably had Herodotus, 2, 35 in mind when he wrote these lines.

properly fed. The healthy outdoor life and hard work made the lower class women sturdy and aggressive. The poor men were bossed unmercifully. Being more muscular than the men, these robust women did as they pleased with them. All their indignities at last got under the men's skins, but it was not the painted darlings of the upper class women who started the rebellion: it was the virile foreign traders who resented having to transact all their business with arrogant females. To their amazement, the tame males, sometime during the fifth century B.C., learned that in other countries it was the men who gave the orders and the women who obeyed. History has not preserved the name of the first hero who hurled a cooking pot at his wife's head. The rebellion of the males followed; the men changed places with the women, and all became as the gods no doubt had intended it to be in the beginning. But the women retained the right of keeping all the property to themselves and willing it to their daughters or other female next of kin. Certain respectable womanly professions remained open to them, because they were more skilful than men at some things, such as midwifery.

Sanitation was highly developed, probably as the hard-won experience of ages of fighting filth and disease. But worms cursed them then as now. On the theory that seemly things should be done in public and unseemly things in private, the Egyptians ate outdoors and, like us, relieved their natural wants indoors. Cleanliness before prudery was their creed, and for that reason they practiced circumcision—one of the customary obligations of the orthodox Jews to this day—which the Egyptians seem to have been among the first to prescribe. They also had learned the hygenic advantages of scoured dishes and laundered linen. They drank from bronze cups, cleansed daily, and washed their clothes every other day. Men, except the priests, wore two garments, women only one. The priests

63

particularly made a cult of cleanliness. To avoid contaminating the gods with lice and "all else foul" that might infest their skins, the supersanitary priests shaved their whole bodies every third day—quite a contrast to the hairy Babylonians. (Ancient history crawls with lice.) In their striving for an almost surgical sterility the priests wore in addition to their single linen garment only clean sandals of papyrus. What they sacrificed to comfort in the name of holiness was more than made up by the abundant fare they enjoyed at the expense of their sweating supporters. One tabu of their diet is historically suggestive when we remember all the hard things Pythagoras (sixth century B.C.) and his disciples said about the lowly though nourishing bean. The priests could not endure even to look at beans, considering them an unclean sort of pulse. Pythagoras shunned beans because—among other reasons—he imagined his dead friends blasphemed in their voice. Another tabu of the Egyptians is familiar to us from its survival in our Old Testament. If an Egyptian gentlemen accidentally touched a hog he rushed to the river for a dip and plunged in, clothes and all. Yet the Egyptians ate pork. The necessary butchering was done by the lowest caste. These pariahs also attended to the more repulsive but unavoidable details of preparing the dead for mummification. Something of the Egyptians' dislike of butchers and embalmers has survived in American civilization: an astute counsel for the defense in a murder trial will accept neither a butcher nor an undertaker on the jury, nor even a doctor.

The priesthood was an exclusively male organization, and no woman had to become a dedicated prostitute. This contrasts with the later Greek custom of employing hierodules in the temples. The women suffered from a heavier disability, however; daughters were compelled to support their parents whether they wished to or not, while sons had no such obligation. A revival of this custom has been attempted in Hollywood in our time when the courts have tried to force successful movie actresses to pay the bills of their spendthrift mothers and fathers.

Although the next has nothing to do with the possible origins of mathematics, I tell it because of the light it throws on the softer side of the Egyptian character and because some famous mathematicians have preferred cats to dogs. From their art it is obvious that the Egyptians appreciated and respected cats. They used them not only to keep down the mouse and rat populations in their granaries, but also as expert assistants in fishing. (Cats hate water only when it is dumped on them, as who wouldn't? Given the opportunity, they will fish diligently in the neighborhood fish ponds.) The Egyptians mummified their dead cats, and in our own nineteenth century shiploads of cat mummies were exhumed and sold abroad for fertilizer. This seems to any respecter of cats more of a desecration of the past than the exhibition of a king's shriveled corpse for morbid tourists to gape at. Whatever the right of this, Herodotus reports that the Egyptians were the only people of his day who kept animals in their houses—especially cats and dogs. He says there would be more household animals if it were not for what happens to the cats. He then describes a cat custom that anyone can observe today: "the tomcats do away with the kittens if they can get at them, but do not eat them." Deprived of their offspring "and desiring to have more," the mother cats take up with the males again, "for they are creatures that love offspring." When a fire broke out the Egyptians thought more of rescuing the cats than of putting out the fire. The cats however did not always wish to be rescued, and would spring over the men attempting to keep them from the fire and leap into the flames. Whether the like ever happens today seems doubtful, but the similar panic of horses in a fire is common. When a cat died a natural death its

65

owners shaved only their eyebrows as a sign of mourning, but when a dog died they shaved from crown to toe. The cats however got the greater honor in the end. For one representation of anything resembling a dog in Egyptian art there must be at least twenty glorifying a cat.

We noted the socialized medicine of the Babylonians. The Egyptians went to the other extreme. The country teemed with physicians, all rigidly specialized and members of the E.M.A. —Egyptian Medical Association. A doctor treated only diseases of the eye, or of the belly, or "hidden diseases," and so on. The populace as a whole seem to have been food faddists, believing that faulty diet was the cause of all sicknesses. They never seem to have suspected the Nile water as the source of their endemic plague of worms. Three days of every month were set aside for purges and emetics. It must have been quite an occasion. The diet was varied enough for any vitamin enthusiast—whole-meal bread kneaded, incidentally, with the feet, sun-dried or brine-pickled fish, small birds eaten either raw or salted, dried lotus lilies, baked loaves of the crushed "poppylike center of the lotus," the sweetish root of the lotus "round and as big as an apple," also seeds, "the size of an olive pit," of another lily with a "calyx like the comb made by wasps." The last is as graphic a description of the plant Herodotus mentions as anyone who has seen the "lily" could ask. Another detail of the health program brings us abreast of today. The river and its marshes made the Nile Valley a gnat's picnic ground. To get any sleep the Egyptians used equivalents of our mosquito nets. On the whole the people seem to have been healthy.

What did these more or less happy people have to do with mathematics? Almost nothing. After the men revolted against the women, they lived and worked largely outdoors in the

Though the Egyptians did little in mathematics because they had no obvious need for more, they left us a picture of how to enjoy life. Their love for flowers and trees was a passion; the very difficulty of growing them in Egypt made them doubly precious. Some of their most beautiful jewelry commemorates flowers. They loved natural beauty, both human and other. Their civilization lived long and, on the whole, happily.

3

Philosophical Interlude

Numbers seem to have had a perverse fascination for philosophers ever since Pythagoras in the sixth century B.C. asserted that "Everything is number" and meant exactly what he said. His discovery that the pitch of a note emitted by a tensed string when plucked depends in a simple way only on the length of the string, was responsible for his all-inclusive generalization. To appreciate Pythagorean arithmetic we must see how this came about.

The dependence of pitch upon length was expressed in terms of whole numbers, or rather their ratios. The *ratio a:b* of the whole numbers a, b is the fraction a/b, as we usually write it today. For a reason that will appear immediately, Pythagoras was interested in the particular ratios, $2:1 = 2/1 = 2$; $3:2 = 3/2 = 1\frac{1}{2}$; $4:3 = 4/3 = 1\frac{1}{3}$; $9:8 = 9/8 = 1\frac{1}{8}$. He observed that when he plucked one string twice the length of another, the second string sounded the note an octave lower than that emitted by the first. Thus the musical "interval" of the octave is 2:1. When he plucked a string $1\frac{1}{2}$ times the length of the first, he got the interval which we call the fifth; a string $1\frac{1}{3}$

times the length of the first gave him the interval we call the fourth; to the ratio 9/8 corresponds the interval between the fourth and the fifth. These experimentally determined facts revealed an intimate and unsuspected connection between the physical universe and whole numbers, the first of its innumerable kind to be recorded in the long history of the physical sciences. Pythagoras' beautifully simple discovery seemed so mysterious to him that he imagined he had found the innermost secret of the universe in whole numbers and their ratios: *everything* is a whole number or the ratio of two whole numbers. Either he or one of his disciples soon proved, however, that the length of the diagonal of a square whose side is one unit long, $\sqrt{2}$, is neither a whole number nor the ratio of any two whole numbers. All this is in any elementary physics course; and any teacher of music knows it without having to think about it. In short it is trite and commonplace. But it is neither trite nor commonplace that *numbers* should govern at least a small part of the knowable universe. It shook superstition to its foundations and prepared the way for the religion generally accepted today.

The shattering disaster of $\sqrt{2}$ was ignored for centuries by a host of wilful believers applying "Everything is number" to such diverse matters as marriage, astronomy and, literally, the "soul" of the universe. The outcome is variously called number mysticism, the philosophy of numbers, and numerology. The last is preferable to the others, as it implies no slur on either mysticism or philosophy. Numerology, according to the dictionary, is "the study of the occult significance of numbers." To make this quite clear, I transcribe what the dictionary says about *occult* (adjective):

Dealing with or in the arts or practices involving the supposed action or influence of supernatural agencies or some secret knowl-

71

edge of them, as alchemy, magic, astrology; also relating to these arts; as *occult* sciences or philosophies.

A second meaning is:

not to be apprehended or understood; beyond the scope of plain understanding; mysterious, supernormal, or supernatural.

As examples of the last are the assertions of the Pythagoreans that odd numbers are male and even numbers female, and that justice is the number four. From Pythagoras to the present, numerology has flourished at one time or another in philosophy, fortune-telling, cosmology, cosmogony, theology and atomic physics.[1]

Mathematicians deny that numbers have any occult significance at all. So for them numerology is vacuous. But any mathematician interested in the historical development of his subject will grant that numerology has incidentally suggested problems of lasting mathematical interest in the theory of numbers. We shall see a spectacular example at the end of this chapter in Plato's curious numerology of marriage. In spite of many such lucky hits, a mastery of numerology can hardly be recommended today as a suitable preparation for a serious study of the theory of numbers.

We are interested here in Pythagoras and his cult, the Pythagoreans, only because of their influence on the theory of numbers. Thousands of books and articles on the life and

---

[1]For an account of this see my book, *The Magic of Numbers*. The book also discusses the lives and works of the major numerologists from Pythagoras and Plato to Eddington (died 1944). For the historicity of the Pythagorean legend, see Chapter 7, first part.

teachings of Pythagoras have been written in the past 2500 years—I once saw an incomplete bibliography listing over 45,000 titles. Unfortunately only one rather spiteful remark by "the weeping philosopher" Heraclitus (flourished 500 B.C.) censuring Pythagoras for allegedly practising "bad arts" is anywhere near contemporaneous, and we have to trust accounts written from about two to five centuries after Pythagoras had passed on to his next incarnation after the human. This applies in particular to two important items in the theory of numbers, perfect numbers in the sense already explained, and "amicable" or "friendly" numbers, both of which occupied Fermat and his contemporaries in the seventeenth century. Both are attributed to the Pythagoreans on, at best, hearsay evidence. I shall defer further discussion of them till a later chapter, where they appear in the documented historical record. For the so-called polygonal numbers there is better evidence, so I shall describe them later in the present chapter.

The legend of Pythagoras, son of the lapidary Mnesarchus, asserts that he was born about 579 (or 570) B.C. on the island of Samos and died about 500, possibly at Crotona, possibly at Metapontum, both in Southern Italy or Magna Graecia. He was inspired to study mathematics and philosophy by the aged Thales (640?–546 B.C.), first of the Seven Wise Men of Greece, whom he visited in Miletus. Thales usually is credited on doubtful evidence with a few simple theorems in elementary geometry, such as that the angle inscribed in a semicircle is a right angle, which he is said to have proved. If he did prove these theorems, he was the inventor of deductive reasoning, at least as applied to geometry. However, the actual development of the deductive method is usually ascribed, by Aristotle among others, to the Pythagoreans of the sixth and fifth centuries B.C. Progress was astonishingly rapid. In only about a century most of plane geometry as studied today in a first school course had been worked out. The invention and application of deductive reasoning was one of the two most important contributions of the Pythagoreans (or of anyone else) to the progress of civilization and its possible destruction, the other being the experimental method ascribed by tradition to Pythagoras alone. Of course there had been experiments before Pythagoras. In the scientific sense of experimentation, definite questions are proposed to be tested experimentally; it is not haphazard, as was the prescientific. Pythagoras' success with the physics of music suggested that nature might be taken out of the hands of capricious and incompetent gods and brought under the faltering but surer control of men.

Having learned from the East himself, Thales urged the young Pythagoras to go to Egypt and Babylonia for instruction from the priests. For personal reasons Pythagoras was eager to go. On his return from Miletus to Samos he had ventured to argue with Polycrates, the able and ruffianly tyrant of Samos, and had rashly won the argument. It was prudent to leave. Tradition credits Pythagoras with twenty-two years of travel and study, mostly in Egypt and Babylonia. Some say he got as far as India. The times were troubled—the dominant and sadistic Persians made tourist travel hazardous. Among their other customs the Persians were addicted to wholesale crucifixion for minor infractions of military discipline. Seeking a comparatively quiet spot to settle in after his wanderings, Pythagoras on his return from the East chose Croton in Southern Italy, at the moment inclined to peace. Most of the other important cities of Southern Italy were enthusiastically pursuing war in its more brutal forms, and so were unsuitable for the philosophic calm Pythagoras sought. He set himself up as a professional "lover of wisdom"—philosopher (he coined the word)—and rapidly acquired an uncritical following of loyal disciples, both men and women. These devoted souls formed the nucleus of the cult later known as the Pythagoreans, or the Pythagorean Brotherhood. Both mathematics and numerology were passionately cultivated by the brothers and sisters. Their life was austere, not to say harsh and drab. Even the satisfaction of credit for their own mathematical discoveries was denied the subservient disciples, and either voluntarily or by discipline they ascribed their personal findings to the Master, Pythagoras.

The Brotherhood seems to have taken the Master's esoteric religion more seriously than any of his suggestive science and sane mathematics. They faithfully practiced the Master's teachings in their daily lives, even to refraining from beans which, until Pythagoras reformed them, they had enjoyed with the cows. Then there were the numerous tabus, some of which seemed to make little if any sense, but had to be observed in the daily ritual—not to sit on a quart pot (who could?), not to stir the fire with an iron poker, and so on for many more. One tabu almost tempts us to believe that Pythagoras foresaw the murderous automobile: don't walk on highways. Another—don't let swallows share your roof—may echo the voice of experience, as anyone who has ever scooped the bugs out of a swallow's nest will agree. The central article of the Pythagoreans' religion, however, was in none of these comparatively sensible precepts, but in the mystical doctrine of the transmigration of souls, possibly imported from Egypt. The Master himself claimed remembrance of many incarnations, clear back to the earliest of all when, as the god Apollo, he invented music. He also asserted that he was the son of Apollo. There is nothing incongruous in this contradiction of habitual common sense. It merely shows that a very great mathematician and an equally great philosopher can on occasion reconcile any contradictions, and can say and believe almost anything.

75

The Pythagoreans came to a bad end through nobody's fault but their own, as in a classical Greek tragedy. Pride, or what in their case was the same, snobbishness, was their undoing. That their pride was not in material possessions but in spiritual intangibles made it particularly offensive to the solider and stupider citizens of Croton, some of whom had been black-balled for membership in the exclusive and intellectual Brotherhood. The despised and rejected swore to get even with the arrogant and supercilious snobs. They did. Democracy (classical Greek model, supported by slavery) had been stirring for some time in Croton when Pythagoras began meddling in civic affairs, always a foolhardy thing for a philosopher or a mathematician to do. He insulted the noisy democrats by exhorting them to clean up their dirty politics. The outraged democrats retorted by setting fire one evening to the lodge where Pythagoras and his closest disciples had gathered for meditation. Some say Pythagoras died in the flames. Others say he escaped to Metapontum, where he spent his last few years. On reaching the age of seventy he suffered an acute attack of disgust with human beings and, hoping for better luck as a dog or a cat, starved himself into his next incarnation.

Greek interest in the properties of numbers began with the mystical Pythagoreans. Their predilection for numerology and geometry held back both them and their earlier successors from a scientific study of numbers. Euclid in the fourth century B.C. gave the first recorded proofs of extremely simple and equally basic theorems concerning numbers, particularly those relating to divisibility. Before Euclid the substitute for proof was inconclusive verification from numerical examples, on an infinitely lower level of difficulty.

For example, Pythagoras is credited by Proclus (fifth century A.D.) with knowing that $x = 2a + 1$, $y = 2a^2 + 2a$, $z = 2a^2 +$

$2a + 1$, where a is any integer, give a solution of $x^2 + y^2 = z^2$. For $a = 1$ these generate the so-called cosmic triangle $3^2 + 4^2 = 5^2$ of Pythagoras which, as we have seen, was known to the Babylonians. The verification is immediate, but no record of the method by which the result was discovered has survived. There are conjectures of course, as always where nothing is known. Likewise for the neater solution $x = a^2 - 1$, $y = 2a$, $z = a^2 + 1$ attributed to Plato. For $a = 2$ this gives the cosmic triangle.

The real problem here for a modern mathematician would be to find *all* the integer solutions. The complete solution was probably beyond the resources of the Pythagoreans, but Euclid was far enough advanced to have obtained it and proved its completeness had he wished. (He gave a solution which includes those of the Pythagoreans and Plato, but did not discuss its completeness.) The verification of Plato's solution, $(a^2 - 1)^2 + (2a)^2 = a^4 + 2a^2 + 1 = (a^2 + 1)^2$, shows only that his $a^2 - 1$, $2a$, $a^2 + 1$ actually satisfy the equation, that is, the sufficiency of his solution is established. Likewise for Euclid's solution. But the question whether every solution x, y, z necessarily is of the form stated is ignored. This failure to take account of both necessary and sufficient conditions leaves the harder part of the problem unattempted.

All the Greek work concerning the solution of equations in integers or fractions is similarly deficient. So also for that matter is a huge and disorderly mass of post-Greek work down to our own times. Our chief advantage over our predecessors is that we usually recognize the distinction between sufficiency and necessity in an arithmetical demonstration, and know when we have proved something about numbers—where the distinction is vital—and when we have not. *We* in the preceding sentence does not include the hundreds of would-be provers of Fermat's Last Theorem, who frequently undo themselves by confusing necessity and sufficiency, especially in deceptively simple questions, such as those that Euclid disposed of, concerning divisibility for integers. As even experts have been falling into this pit for the past two centuries we need not be too critical of the Greeks.

Always with the reservation that anything about numbers attributed to the Pythagoreans is second- or third-hand at best, we may note some of what they are said to have done. They distinguished between odd and even numbers, and proceeded immediately to the numerology of the distinction: odd numbers are male and good; even numbers, particularly the smallest, 2, are female and bad. They started the long philosophical debate concerning the number 1; was (is) it, or was (is) it not, a "number"? If a number is a collection of units, as some asserted, there was an obvious difficulty. Again, the generation of all numbers from 1, as $2 = 1 + 1$, $3 = 2 + 1 = 1 + 1 + 1$, and so on, was hard to account for if 1 was (is) not a number. Certain pagan philosophers, seconded in the early centuries of Christianity by able theologians, identified 1 with God. This ended the debate. So much for sacred numerology.

On the profane level, the Pythagoreans recognized that even times even is even, as also is even times odd, while odd times odd is odd. The philosophical implications of these truths were as remarkable as they were far-fetched and far-reaching. Going beyond merely even numbers, the Pythagoreans discussed multiples of four, called evenly even numbers, at some length. Such things are trivially easy to us today, but they were not easy to a people who had no universal and usable symbolism for numbers. Some of the Pythagorean terminology in these matters, such as "evenly even," survived well into the nineteenth century.

A more interesting detail was the theory of arithmetic, geometric, and harmonic "means" traditionally ascribed to the

Pythagoreans long after their sect had become extinct. The *arithmetic mean* of two numbers a, b is defined to be one-half their sum, $(a + b)/2$; their *geometric mean* is $\sqrt{ab}$. If in the sequence $a, b, c, d \ldots$ the differences of consecutive terms are equal, so that $b - a = c - b = d - c, \ldots$, the sequence is said to be an *arithmetic progression*. Note that b is the arithmetic mean of the two numbers a, c adjacent to it, $b = (a + c)/2$, likewise $c = (b + d)/2$, and so on all along the sequence. These definitions are to hold for both whole numbers and fractions. If the fractions $1/x, 1/y, 1/z, 1/w, \ldots$ are in arithmetic progression, $x, y, z, w, \ldots$ are defined to be in *harmonic progression*, and y is the *harmonic mean* of x, z; z is the harmonic mean of y, w, and so on. The name goes back to Pythagoras and his plucked strings, the successive fractions in the progression corresponding to the harmonics of a fundamental tone. From the last definition, $1/y - 1/x = 1/z - 1/y$, whence the harmonic mean y of x and z is $2xz/(x + z)$. The arithmetic mean is $(x + z)/2$. From these two means we see the simple fact that astonished and bemused the Pythagoreans: the ratio of the first, x, of two numbers x, z, to their arithmetic mean, $(x + z)/2$, is equal to the ratio of the harmonic mean, $2xz/(x + z)$, to the second number, z. Anyone who remembers how to manipulate fractions in school algebra will dismiss this somewhat involved statement as an obvious triviality. To see that it may not have been evident to the Pythagoreans, who knew nothing of algebra, try to prove it in words directly from the definitions of the means concerned. As a numerical example of great historical interest, the arithmetic mean of 12 and 6 is 9, the harmonic mean is 8, and

$$12:9 = 8:6.$$

This is a very famous relation in Pythagorean arithmetic

and music, also in Platonic numerology as in the *Timaeus*, and is said to have been brought from Babylon to Croton by Pythagoras. It was called the perfect proportion—the numbers a, b, c, d are defined to be *in proportion* if $a:b = c:d$ or, as we usually write it, if $a/b = c/d$, or what is equivalent, if $ad = bc$. The numbers a, b, c, d are called the *terms* of the proportion. To us it is trivial that a proportion is not destroyed if all its terms are multiplied or divided by the same number. From the perfect proportion, for example, with 6 as divisor we get the following equivalent proportion,

$$2 : \tfrac{3}{2} = \tfrac{4}{3} : 1,$$

in which we recognize the musical intervals discovered by Pythagoras. The harmonies implicit in this simple relation between numbers were responsible for some of Plato's most remarkable science and philosophy. Another numerical fact of the same sort also inspired much abstruse philosophical speculation on the constitution of the universe: if A, G, H are the arithmetic, the geometric, and the harmonic mean of any two numbers a, b, whole or fractional, G is the geometric mean of A and H. I refer the reader to the *Timaeus* for the relevant numerology.

I return for a moment to the equation $x^2 + y^2 = z^2$. The first proof of the geometrical interpretation of this—the square on the longest side, z, of a right-angled triangle whose shorter sides, or "legs" x, y, contain the right angle is equal to the sum of the squares on x and y—is traditionally credited to Pythagoras himself. The figure of the right triangle suggested innumerable things to do geometrically and then to interpret in terms of numbers. Two examples will be enough. Drop a perpendicular from the right angle to the longest side. For what integers x, y, z will the length of this perpendicular be an integer? Or for what integers x, y, z will the area of the triangle be an integer? Or the square of an integer? But after all a right triangle is a very special kind of triangle. Why stop with it, instead of going on to any triangle? We might ask for what integer sides is the area of any triangle an integer? This was *completely* answered only in the present century. Then we might complicate the triangle by including the lines (the medians) drawn from the vertices to the mid-points of the opposite sides, and ask for what triangles with integer sides the medians are also integers. And so on, even to figures other than triangles. To make the questions harder, we might require the lengths of certain lines to be squares or cubes of integers. This seductive diversion occupied mathematicians off and on for centuries. There was an orgy of it in the seventeenth and eighteenth centuries, and the nineteenth also indulged, but not quite so incontinently. Then, in the nineteenth century, the geometric phraseology was gradually relinquished, and problems that Fermat and his contemporaries stated—or might have stated—in the language of geometry were restated algebraically with no reference to any possible geometrical origin or representation. For instance, the problem of finding right triangles with integer sides and integer area equal to twice a square is equivalent to that of solving the simultaneous equations

$$x^2 + y^2 = z^2, \qquad xy = 4w^2$$

in integers x, y, z, w. If for any such problem there are no solutions, this is to be proved; if there are solutions, all are to be found. The outcome of all this is a vast and constantly growing branch of the theory of numbers, called the arithmetical theory of forms. Anyone but an educated mathematician may be excused for imagining that mathematicians in constructing this theory have frittered away their lives in the elaboration of an intricate futility. I shall not take time out now to indicate why such an opinion is mistaken, even on what may be called the level of practical utility, but may do so in the concluding chapter.

If Pythagoras was indeed the first to prove that the square on the longest side of a right triangle is equal to the sum of the squares on the other two sides, there is no record of his proof. Dozens of proofs have been constructed, some intuitive enough to have been noticed by the earliest geometers who attempted to prove anything. The proof usually given in school geometries is from Euclid's *Elements* (about 330 B.C.), and is supposed to have been Euclid's own. Legend says that Pythagoras was so elated when he invented his own proof that he sacrificed an ox to the immortal gods of his time. Improving on this, some say he sacrificed a hecatomb—a round hundred of oxen. Plutarch (46?–?120 A.D.) in his *Moralia* cites the more modest version to illustrate the superiority of intellectual satisfactions over grosser and more popular pleasures:

Though a man with a voluptuous temperament may be naturally eccentric and rash, no such ever sacrificed an ox for joy when he had gained the favor of his mistress. But Pythagoras sacrificed an ox when he had proved his theorem.

Today in similar circumstances a mathematician sacrifices no more than a magnum of champagne, if he has the price.

The Pythagoreans are said to have invented the *polygonal numbers*, which enumerate the pebbles or dots arranged in the form of equilateral triangles, squares, regular pentagons, regular hexagons, regular heptagons, and so on. The general idea is clear from the first three kinds. A polygon (triangle included) all of whose sides are equal and all of whose angles are equal is said to be *regular*. The *triangular numbers* refer to the regular (equilateral) triangles of $1, 3, 6, 10, \ldots$. dots as in the figure (1 is included as the first in all kinds of polygonal numbers by convention);

the next, 15, is obtained by the same kind of bordering from the 10, by adjoining *one* row (of 5 dots).

Counting the dots in the successive rows of the last we get

$$1 + 2 + 3 + 4 + 5 = 15,$$

and likewise for $1, 3, 6, 10$, for example

$$1 + 2 + 3 + 4 = 10.$$

The next after 15 is obtained by bordering 15 with *one* row (of 6 dots),

$$1 + 2 + 3 + 4 + 5 + 6 = 21.$$

The general rule is apparent: the nth triangular number is

$$1 + 2 + 3 + \cdots + n.$$

83

I leave it to the reader to verify that this is $\frac{1}{2}n(n+1)$. (Write the numbers $1, 2, 3, \dots, n$ in the reverse order, $n, n-1, \dots, 3, 2, 1$, and add the 1st, 2nd, 3rd, ... numbers in these two sequences.)

The *square numbers* are represented by

or $1, 4, 9, 16, \dots$. The next, 25, is obtained by bordering the 16 as below,

and this kind of bordering on *two* adjacent sides applies all along, after 1,

From this we get

$$4 = 1 + 3, \qquad 9 = 4 + 5 = 1 + 3 + 5,$$
$$16 = 9 + 7 = 1 + 3 + 5 + 7,$$

and generally,

$$n^2 = 1 + 3 + 5 + 7 + \cdots + (2n-1).$$

The proof of this is frequently set as an exercise in the school algebras in the chapter on mathematical induction.

The *pentagonal numbers*, represented by dots arranged to form a square with a triangle on top,

are $1, 5, 12, 22, 35, \ldots$ and are obtained by successive borderings along *three* sides. The first ten triangles, ..., heptagons are as in the table, equivalent to that given in Greek symbols by Nicomachus in the first century A.D. (We shall meet him later.)

Figures	Numbers
Triangles	1, 3, 6, 10, 15, 21, 28, 36, 45, 55
Squares	1, 4, 9, 16, 25, 36, 49, 64, 81, 100
Pentagons	1, 5, 12, 22, 35, 51, 70, 92, 117, 145
Hexagons	1, 6, 15, 28, 45, 66, 91, 120, 153, 190
Heptagons	1, 7, 18, 34, 55, 81, 112, 148, 189, 235

The reader may be interested in the formula for the rth polygonal number with n sides,

$$\frac{r}{2}[2 + (n - 2)(r - 1)].$$

Thus for the 5th heptagonal number, $r = 5$, $n = 7$, and the required number works out as 55. For $n = 3$, we check that the rth triangular number is $r(r + 1)/2$; for $n = 4$, the rth square number checks as r^2.

85

The Pythagoreans and their successors, before the development of a usable algebraic notation in the sixteenth and seventeenth centuries of our era, inferred many interesting facts from dot representation of the numbers as above. If continued interest in a mathematical topic is a measure of its importance, the polygonal numbers were the most important contribution of the Pythagoreans to the theory of numbers. Fermat and his correspondents in the seventeenth century investigated the properties of these numbers intensively, starting several lines of work that continued through the eighteenth and nineteenth centuries and well into the twentieth. With advancing knowledge and improving mathematical technique, more and more difficult problems were proposed and attacked with varying success.

A not-too-easy problem presented itself early to Diophantus, who, apparently, gave it up. In how many ways is a given number, say 6, a polygonal number? From the short table of Nicomachus, we see that 6 is both a triangular and a hexagonal number, and a moment's thought (confirmed by the table) will make it clear that no other polygonal number is 6. If the table had extended we should have noticed that 36 is the second 36-sided number, the third 13-sided number, the sixth 4-sided number (square), and the eighth 3-sided number (triangle), and no more. So 36 is a polygonal number in exactly four ways. The problem is to find the number of ways in which any given number is polygonal without constructing tables. Anyone who tries to solve it explicitly and completely will see why it stopped Diophantus. If, as is now sometimes supposed, he lived in the first century A.D.—or even in the third—he could not possibly have got through to the end with what he knew. Nevertheless he tried and took the first long step.

As an indication of the continued interest in polygonal numbers from the Pythagoreans to the twentieth century, Dickson's *History of the Theory of Numbers* reports on 282 papers dealing with them down through the year 1919, and since then there have been several more. Why all this activity? The previous remarks on the arithmetical theory of forms apply here. And a subject that has in one form or another attracted mathematicians of the caliber of Diophantus, Fermat, Euler, Legendre, Cauchy and Gauss can hardly be dismissed as worthless or trivial.

To finish with the polygonal numbers, I take two long steps ahead, out of the chronological order, to the seventeenth and nineteenth centuries, and state a capital theorem highly prized by arithmeticians for its simplicity and completeness.

To avoid needless repetitions, I follow the usual custom of adjoining zero to the rational integers, so that we now consider

(A) $0, 1, 2, 3, 4, \ldots$

instead of

(B) $1, 2, 3, 4, \ldots$.

For example, the first instance of the general theorem can be stated thus: every integer is a triangular number, or a sum of two such numbers, or a sum of three such if we refer to (B); if we refer to (A), we state inclusively that every integer is a sum of three triangular numbers, understanding that some of these numbers may be zeros, and so for squares, pentagonal numbers, etc. Also, instead of triangular numbers, square numbers,..., pentagonal numbers,..., I shall say 3-sided, 4-sided, 5-sided,..., n-sided numbers, thus discarding the antiquated historical names, such as the Greek equivalent, whatever it may be, for 123477-sided numbers.

Fermat stated (1636), as usual without proof, that every integer is a sum of n n-sided numbers, for all n. He says:

I was the first to discover [this] very beautiful and entirely general theorem .... I can not give the proof here, which depends

87

upon numerous and abstruse mysteries of numbers; for I intend to devote an entire book to this subject and to effect in this part of arithmetic astonishing advances over the previously known limits.[2]

A. L. Cauchy (1789–1857) gave the first proof in 1813–15, repeated in 1826, In the latter, the proof fills thirty-one printed pages.

So far as we know, the Pythagoreans were not interested in calculation. The Babylonians and the Egyptians, as we have seen, were. It may have been the aloof Pythagoreans who set the Greek fashion of slighting calculation in favor of investigating the properties of numbers as such. Nevertheless, the Greek astronomers could not have got on without calculation, and some of them, notably Ptolemy in the first century A.D., were highly skilled calculators. The Babylonian sexagesimal notation was a great help in actual computation. But the mathematicians, as opposed to the astronomers, busied themselves almost exclusively with properties of numbers having no apparent relevance for calculation. This kind of theoretical work with numbers was called *arithmetica*; calculation was called logistica, or *logistic*. Some of the Greek intellectuals, notably the Sophists, did not despise logistic and in fact cultivated it, but they were a small minority. Arithmetica was the Greek precursor of our theory of numbers. The oppressive authority of Plato, as a philosopher might expect, was wholly on the side of arithmetica. It will be interesting to see shortly why this may have been so.

---

[2] Dickson's translation, *History of the Theory of Numbers* (Washington, D.C.: Carnegie Institute, 1920), II, 6. The promised book never appeared. Note that the date, 1636, is a year earlier than the statement of Fermat's Last Theorem. Fermat was in his early thirties, his great period.

I must first give a few facts of Plato's life. It will be illuminating to note that Plato (427?–347 B.C.) claimed descent from the first king of Athens. On both sides he was related to prominent aristocratic families of Athens, and received the customary education of a youth in his distinguished station. It was not a very usable education, and whatever science and mathematics it may have offered was reduced to the literary minimum of dilettantism. Plato's amateurish interest in mathematics was to come late when, as a young man, he set out on his travels to escape the political turmoil in Athens and to complete his education. There is a legend that he learned some mathematics from the priests of Egypt. But his real teachers were members of the suppressed and rapidly dwindling Pythagorean sect. He became the most influential exponent of the Pythagorean philosophy and numerology in all its long history. Early in his travels he heard of the "bible" of Pythagoreanism by Philolaus of Thebes (flourished in the late fifth century B.C.) and, according to tradition, paid a high price for a copy. Whether he actually acquired this invaluable and mystifying work is unimportant. Wherever he may have found the Pythagorean teachings, he thoroughly assimilated them. Second only to his interest in Pythagoreanism was his passion for political theory. He was never a conspicuously successful politician, being too academically naïve, and too eagerly credulous of the ultimate perfectibility of human nature, to understand the practical game of politics as it is played by astute dictators and shrewd petty bosses. He also had the handicap of an effeminate and squeaky voice. His theory of good government in the celestial blue is the theme of one of his most celebrated dialogues, the *Republic*. At the invitation of Dionysius the Elder, tyrant of Syracuse, he undertook to reform that thoroughly rotten but happy community and its

89

dissolute rulers, making in all three attempts, two on the tyrant himself and one on the genial ruffian's heir. Plato was approaching middle life at the time, and had acquired an amateur's enthusiasm for the more philosophical aspects of elementary geometry. He had a typical philosopher's holiday. In the simplicity of his soul he tried to redeem the wily tyrant by tutoring him in the mysteries of triangles and rhombuses. The lessons did not accomplish the purposed end, and Plato returned to Athens with expressions of gratitude and respect from the tyrant. He has learned little of political human nature. Subsequent excursions to Syracuse also ended in nothing but unwanted gifts and insincere compliments from the unregenerated rulers.

But Plato did not lose his faith in the perfectibility of the human race as it was in his day. The Peloponnesian War (431–404 B.C.) in which Sparta humiliated Athens is said to have inspired him to work out the utopian theory of government retailed in his *Republic*. The seething corruption of Athenian politics and the cultured softness of the Athenian people, especially the effeminacy of the young men, contrasted with their opposite numbers in Sparta, were partly responsible for Plato's ideal politics. The Spartans, as virile as gorillas and as hard (including their heads) as bricks, provided a first crude model from which to proceed by mathematics and philosophy to ethereal perfection. Sparta was the complete totalitarian society, where the rule was one for all and all for one, the one being the state. Plato's emasculated version in his *Republic* still left the mass of mankind at the bottom of the social heap. The problem was to provide a workable system of government in which the distribution of consumers' goods would remain as it was, and criticism of things as they were would be discouraged as subversive and blasphemy against the established gods. The purpose of these measures was, of course, asserted to be, as usual, the government of the governed for their own temporal and eternal good. The governed did not always agree—even as you and I.

At the age of forty Plato settled for the rest of his long life in Athens. He founded his famous Academy for the advancement of science (Pythagorean), mathematics (also on the Pythagorean side), and his own comprehensive philosophy. He had numerous students, some very earnest and others slightly confused in their determination to master the Platonic dialectic. In his old age, or perhaps dotage, he proposed his theory of "ideal numbers" to rationalize the Pythagorean "Everything is number." This has nothing in common with the theory of numbers as it is understood today, so we need not consider it here,[3] although it continues to inspire theses for the Ph.D. degree in philosophy.

A famous saying attributed to Plato (it is not in his works) asserts that "God ever geometrizes," or "God is always doing geometry," and Plutarch in his *Moralia* attempts to explain why Plato may have said such a thing. Part of what Plutarch describes suggests why Plato preferred arithmetica to logistic and tells us what Lycurgus (396?–?323 B.C.), the reputed law-giver of Sparta, thought of logistic as a training for youth. The setting is an after-dinner discussion between Plutarch and

---

[3]There is a curious parallel, little if any better than a rather poor mathematical pun, between Plato's assertion that "ideal numbers cannot be added," and the same today in the theory of algebraic numbers—which evolved from attempts to prove Fermat's Last Theorem. However, by a rather artificial definition of *addition*, it is possible to add ideals, as defined in Dedekind's theory.

three friends who give their opinions of what Plato meant. Plutarch sums up when the others have spoken.

To the first speaker everything is clear. Plato, he asserts, held that the true purpose of geometry is to lead us from the sensory and perishable toward the spiritual and eternal, and contemplation of the eternal, he declares, is the end of philosophy. The diagrams of the geometers are irrelevant for the eternal "form" or "idea" of a triangle, a rhombus, and so on. It is with these perfect and unchangeable ideas behind mere mathematical geometry that God is said to be necessarily occupied. Otherwise there would be no mathematical triangles or the rest for geometers to discuss.

The second speaker begins by asserting that Plato linked Socrates with Lycurgus, and considered the traditional author of the Spartan constitution as equally important for Socrates as Plato acknowledged Pythagoras to be for himself. Developing this alleged influence of Lycurgus on Socrates, the speaker, evidently drawing on the theory of government as developed by Plato in his *Republic*, states that Lycurgus banished the study of common arithmetic (logistic) from Sparta. Why? Because the effect of arithmetic is democratic and popular, in the derogatory sense of pulling the aristocracy down and the proletariat up. Such a leveling was undesirable to those at the top, especially to those called "guardians" in the *Republic*. To avoid it, Lycurgus introduced the study of geometry, because in his opinion it was better suited than arithmetic to a sober oligarchy and a sound constitutional monarchy. For arithmetic, by its use of number distributes goods equally, whereas geometry, using proportion, allocates goods according to merit. Geometry therefore is not a source of confusion to the state. On the contrary, it provides the grounds for an infallible means of distinguishing good men from bad. The good and the bad are not given their shares according to mere numbers or lot, but by judgments based on virtue and vice. This desirable end is not attainable by counting, especially of ballots, or any other chicanery of democratic political arithmetic. Nemesis (distribution of what is due) and Dike (custom) should teach us to regard justice as equality but not equality as justice. God attains the just end by the geometrical system of proportion applied to human affairs. He protects and maintains the distribution of things according to merit, regulating it geometrically, that is, in accordance with proportion and law. From this, Lycurgus concluded that arithmetic is bad medicine for the state, while geometry will preserve it in perfect health and lusty longevity. Such arguments today are nonsense, but in Plutarch's time they may have made sense.

The third speaker expounds from another of Plato's platforms. This one was set up by the Pythagoreans. Geometry imposes order and harmony on matter, which is the principle of disorder and discord. By putting number and proportion into matter the unlimited is bounded and circumscribed, first by lines, then by surfaces, and finally by volumes. In this way the first forms and different bodily shapes condense and come into being. These primal forms embody the four classical "elements"—air, water, fire, and earth. The good word this speaker has for "number" refers to the eternal idea of a number, and not to what Lycurgus and Plato disliked.

Plutarch, evidently drunker than his guests, summing up, agrees in principle like a polite host with what the others have said, and adds an interesting obituary on that unfortunate ox Pythagoras sacrificed. The function of geometry, he agrees, is to elevate the mind from earthly to celestial things. Politically, arithmetic is democratic and geometry oligarchic, or, as it might be phrased in the 1950s arithmetic is communistic, geometry capitalistic. Cosmically, geometry is the key to unlocking the secrets of the universe. The inborn mathematics of the mind is necessarily superior to sensory experience. As for God the Geometer, Plato meant that God, being the Supreme Geometer, had set himself the hardest and most important problem he could imagine in celestial geometry. This was not imitated by Pythagoras, as might be hastily supposed, in the geometrical form of $a^2 + b^2 = c^2$ (a, b, c being the sides of a right triangle), but in the much finer problem which Pythagoras proposed and solved in its earthly version: to construct a figure equal in area to one given figure bounded by straight lines and similar to another figure bounded by straight lines. (This is Proposition 25 in Book 6 of Euclid's *Elements*.) It was this problem, according to Plutarch, that endowed the ox with immortality at the cost of its mundane life. God's solution of the philosophical equivalent was the cosmos, in which form was imposed on all matter.

To conclude this philosophical interlude, I reproduce a modern interpretation of one of the clearest and profoundest examples of Platonic numerology at its best. The interested reader will find many further suggestive examples in Plato's *Timaeus* and his *Laws*. The first of these timeless masterpieces especially should be consulted to appreciate how deeply Pythagorean arithmetica penetrated and undermined the thinking of the most renowned philosopher of antiquity, and some of that today.

Plato's description of the famous "Nuptial Number" in Book 8 of his *Republic* is such a snarl of what at first reads like pure nonsense, that it has inspired a vast and more or less scholarly literature of criticism and commentary in attempts to unravel its meaning. We need not consider any of these efforts to make sense out of what after all may be only a philosophical befuddlement of arithmetic, since a strictly arithmetical interpretation, by a professional mathematician, is of more interest here, although it may not be what Plato had in mind. Of course nobody, not even a Platonist, can prove that he knows what Plato here was talking about. This is what he says:[4]

For that which, though created, is divine, a recurring period exists, which is embraced by a perfect number. For that which is human, however, by that one for which it first occurs that the increasings of the dominants and the dominated, when they take three spaces and four boundaries making similar and dissimilar and increasing and decreasing, cause all to appear familiar and expressible.

Whose base, modified as four to three, and married to five, three times increased, yields two harmonies: one equal multiplied by equal, a hundred times the same; the other, equal in length to the former, but oblong, a hundred of the numbers upon expressible diameters of five, each diminished by one, or by two if inexpressible, and a hundred cubes of three. This sum now, a geometrical number, is lord over all these affairs, over better and worse births; and when in ignorance of them, the guardians unite the brides and bridegrooms wrongly, the children will not be well-endowed in their constitutions or in their fates.

The outlook for the brides and bridegrooms may not be as dark as Plato feared, for the late Mrs. Young made simple but by no means trivial or obvious arithmetical sense out of all the mysterious specifications for the auspicious numbers. Married to the distinguished mathematician, the late W. H. Young, she

---

[4]Quoted from Grace Chisholm Young (Mrs. W. H. Young), "On the Solution of a pair of simultaneous Diophantine equations connected with the Nuptial Number of Plato," *Proceedings of the London Mathematical Society*, Ser. 2, XXIII (1924–25), 27–44.

was exceptionally well qualified to speak with the authority of intimate personal experience:

Long intercourse with mathematicians has taught me that their covert allusions to mathematics, as throwing light on philosophical or other matters, are usually as profound as their own mathematical knowledge:[5] and some years ago, when I first found myself face to face with the question what mathematical truths Plato was referring to in these oracular utterances, I felt that Plato was here epitomizing some part of his own mathematical cogitations, and that, in unravelling this mystery, we should be gaining a clue by which we may be able to trace somewhat of the intellectual biography of one of the most ancient and eminent of Greek mathematicians [!]. Tradition tells us little about this.

Is it any wonder?

Following some of the more sober commentators, Mrs. Young says Plato is referring to the "Cosmic Triangle," the right triangle whose sides are 3, 4, 5, as his basic fact: 3 being odd, is male, 4 being even is female, and 5 is the Pythagoreans' number for marriage. Also, Plato's "perfect number" is 6, = 1 + 2 + 3, and this is the area of the Cosmic Triangle, which also has something to do with the union of the sexes in Pythagorean numerology. Mrs. Young remarks that "As Plato evidently regarded the degeneracy of the human race as belonging to what we now call Eugenics, it was only to be expected that he would make the Muses begin from the Cosmic Triangle." An almost forgotten commentator, "who lived between Cicero and Ptolemy [the astronomer, second century A.D., not Alexander's general, whom we shall meet later],

---

[5]She must have been more fortunate than some of us.

connects the passage with the equation

$$3^3 + 4^3 + 5^3 = 6^3."$$

This gives 216 ($= 6^3$) for one of the two numbers Plato is said to be describing. The other is supposedly $60^4 = 12960000$, which is strongly reminiscent of the Babylonians with their base 60. Scholars seem to be in fair agreement on these two. If this is all the meat Plato concealed in his untidy package, we can only ask why he went to so much trouble to wrap it up. Mrs. Young replies that Plato slipped in a great deal more:

I fancied that Plato knew, or at least with the intuition of the mathematician, had jumped to the conclusion that *all the solutions possible of the two simultaneous equations*:

$$x^3 + y^3 + z^3 = t^3, \qquad x^2 + y^2 = z^2$$

in integers are of the form

$$3m, 4m, 5m, 6m$$

Anyone who has the persistence to follow Mrs. Young through her difficult and rather lengthly demonstration, will suspect that Plato could not have proved the conclusion he jumped to. Jumping to conclusions in the theory of numbers frequently lands the jumper in a booby trap. If Mrs. Young was right about Plato's "intuition of the mathematician," he was merely luckier than he deserved to be. It is almost a mathematical miracle that he sailed clear over the trap and landed on his feet. Even the incomparable Diophantus, whom we shall meet later, the greatest master of numbers in antiquity, neither proved nor guessed any fact about numbers as deep as what Mrs. Young says Plato's intuition gave him. The following remarks may suggest why.

The equation $x^3 + y^3 + z^3 = t^3$ has an infinity of integer solutions other than the first one $x = 3m$, $y = 4m$, $z = 5m$, $t = 6m$. Not till the late 1940s was the *complete* integer solution of this difficult equation obtained. The added condition $x^2 + y^2 = z^2$ makes the problem much harder, and it is only by the grace of the Great Arithmetician Himself that this condition isolates *just one* set of numbers out of the infinity satisfying the first equation.

Intuition about numbers can sometimes be comic. Many years ago a hopeful aspirant to mathematical fame, seeking to improve on Fermat's assertion that $x^3 + y^3 = t^3$ has no integer solutions, sent Cauchy (1789-1857) his "intuition of the mathematician" that $x^3 + y^3 + z^3 = t^3$ has no integer solutions. Cauchy scribbled $3^3 + 4^3 + 5^3 = 6^3$ on the margin of the manuscript and returned it to its author without comment.

4

Alexander's Contribution

It may well be asked, as one eminent humanistic authority in Hellenic civilization did, what has Alexander the Great[1] (356–323 B.C.) to do with mathematical history, and in particular with Fermat's seventeenth-century France and his famous Last Theorem? Much: without Alexander there might have been no Alexandria and no Ptolemy I (Soter) and his son Philadelphus fostering the Alexandrian Library and Museum, without which there might not have been a golden age of Greek mathematics. There might not have been a recorded Diophantus, and certainly not a Hypatia to edit and transmit what remains of his work on the theory of numbers which, in Bachet's seventeenth-century edition, was to incite Fermat to his greatest work and his still unsolved problem.

Alexander's mother, Olympias, was a wife of King Philip II of Macedon (382–336 B.C.). Philip had eight wives, a retinue of

---

[1]For an enthusiastic appraisal of Alexander as "the greatest man the human race has yet produced," see F. A. Wright, *Alexander the Great*. This extraordinary book reads as if it were the result of a close collaboration among Adolf Hitler, Benito Mussolini and Joseph Stalin.

concubines, and an unquenchable thirst for alcoholic beverages. The Macedonians drank to get soddenly drunk. Before he was twenty, Alexander could drink down any five of his compatriots, and frequently did. He also was an epileptic.

Olympias has been pictured as an aggrieved lady or a neurotic bitch, according to taste. When Philip had all he could take of her, he began making improper advances to a certain Cleopatra (not Shakespeare's lady of the asp, who was still far in the future). After her son Alexander's death, Olympias saw to it that one of his brothers was murdered. But this sort of thing could not go on indefinitely. When she surrendered some years later at the siege of an unimportant town, the commanding general had her butchered. She was a brilliant if rather too imaginative a liar, frequently reporting amorous snakes in her bed, possibly to talk herself out of some indiscretion that might have annoyed her husband. Her favorite snake was a big black fellow who, she asserted, was the Father of Gods and Men in one of his numerous seductive disguises. The fruit of this celestial union, according to Olympias, was Alexander who, therefore, was in some degree divine. When he was old enough to understand what his mother was talking about, he greedily swallowed the genealogy, snake and all. Seventeen years after leaving home at the age of sixteen Alexander had conquered all he knew of the world. On rolling into Babylon on a tremendous baggage truck decked with vine leaves and bunches of grapes, he announced that he was the god Bacchus. He was merely somewhat drunker than usual. The preposterous bosh he had learned at his mother's knee paid off; the vanquished Babylonians pretended to believe him. Did father Philip credit what Olympias told him of Alexander's parentage? Probably not; he was just a simple soldier not given to theology and metaphysics. He eloped with his Cleopatra and left Olympias to her snaky bed. He had other things to occupy him. His life's ambition was at last within his reach. He would invade, conquer, and loot the opulent Persian empire. But he had reckoned without his estranged wife. She hired one Pausanius to murder him and Cleopatra and the latter's young daughter. The conquest of Persia was to be her divine son's greatest triumph—provided his doting mother did not catch him asleep with an ax in her hand. In his progress to glory Alexander surpassed both his father and his mother.

Alexander had a half brother, Ptolemy I, later called Soter (Savior). Philip, as might be expected, fathered several by-blows, of whom Soter was the ablest and most distinguished. His mother was the courtesan (court prostitute) Arsinoë. She had brains as well as beauty, and was a much cleverer woman than the bloody Olympias. Of course it cannot be proved that Philip was Soter's father, but Soter believed that he was and claimed Alexander as his half brother. To silence unpleasant gossip, Olympias declared that "the knight" Lagus, not her dissolute husband, was responsible for Soter—a fairly respectable genealogy. Alexander and Soter grew up together, and even the devoted mother had to admit that Soter was the quicker of the two in everything from gymnastics and swordplay to headwork. When they were ready for school they shared a common tutor, no less a pedagogue than the immortal philosopher Aristotle. Alexander was then thirteen. The boys were unimpressed by their teacher. He lisped and his manner was pedantic. From Aristotle, however, Alexander may have acquired his callous contempt for all barbarians (non-Greeks) as subhuman cattle fit only to be slaughtered. But he made an exception of royal families, even marrying or fornicating himself into them and encouraging his generals to do likewise. Alexander had a broad streak of snobbery in him, and tried to compensate his Macedonian boorishness by toadying to wealthy barbarians and supercilious Athenians.

Philip decided that three years of Aristotle was enough for any young man. Moreover, he was anxious to get on with his war against Byzantium (later Constantinople and Istanbul). Expecting to be gone some time in pursuit of the rich and hated Persians, he appointed Alexander, then only sixteen, regent of the kingdom in his absence. It was a risky thing to do, but possibly safer than leaving Olympias in charge.

Alexander's career after he escaped from Aristotle has been cited by philosophers and humanists as a testimonial to the practical value of an education in philosophy. As nobody knows exactly what Aristotle taught his headstrong pupil, the philosopher cannot be justly debited with Alexander's paranoiac conceit, his sottish drunkenness, his treachery, his sadism, and his ferocious brutality, when irritated, to ally and enemy alike. Against all this, on the credit side may be put the well-substantiated evidence that he grew to detest Aristotle. On a civilized level, the philosopher is said to have been responsible for Alexander's alleged interest in science. It may be true that Alexander when campaigning sent back all sorts of interesting curios and cock-and-bull stories to his old teacher, who is reputed to have had a passion for such things. Whatever interest Alexander may have had in science is partly balanced by his superstition, crass even for his own day. He did take a little of the conceit out of the Greeks by teaching them some geography, some demography and some history at the cost of his bloody campaigns. In return for what they got from the barbarians, the Greeks left at least a futile memory of their own civilization in a rapidly fading dream of "one world" on the Hellenic model.

If philosophy is not to be blamed for Alexander, it is only fair that it should not be praised for his enlightened half brother Soter. However, to give the philosopher what may have been his due, Soter was the first professional soldier on record to put into practice Aristotle's (or Plato's) theory that even the material welfare of a people is furthered by science and philosophy. If someone gets hurt in the process of spreading enlightenment, it is his own fault for being stupid and may be good for his soul.

Only sixteen when Philip rescued him from Aristotle and made him regent of the kingdom, Alexander took on a load of responsibilities such as no other teen-ager has ever shouldered. The kingdom was in turmoil. Alexander quickly settled all disputes between rival factions with the unanswerable argument of the sword. To teach the rebellious Thebans the folly of disobedience and the futility of resistance, he destroyed their capital. At the battle of Chaeronea (338 B.C.) he gave a startling exhibition of the personal courage which was to astonish and overawe the Persian hordes—when he could catch up with them. Whatever may have been Alexander's faults and weaknesses, cowardice was not one of them, nor was shirking dangerous opportunities. When in 336 B.C. Olympias had her husband liquidated, Alexander, not yet twenty, shoved her aside and proclaimed himself king and sole ruler of Macedonia. He then, accompanied by Soter, set out to conquer what he knew of the habitable world. He did not know half of it: he never even saw Italy or Spain, for instance, to mention only two of several countries he did not visit. So his famous (and maudlin) tears over a lack of "more worlds to conquer" might have been spared. At first he was as reasonable and as magnanimous as could be expected of a professional soldier when victory was cheap. But as he passed from one overwhelming conquest to another, he became the classic instance of Lord Acton's dictum that "Power tends to corrupt; absolute power corrupts absolutely."

Having secured his rear, Alexander crossed the Hellespont to Asia Minor (334 B.C.) with an army of 30,000 infantry and 5000 cavalry. Though this sounds impressive, it was negligible beside the Persian hordes Alexander was to rout. As Alexander demonstrated, a well-drilled and intelligently disciplined force of 35,000, competently led, can make a bloody hash of an armed mob of 600,000 or even 1,500,000 commanded by a military imbecile. The first mob was cut to pieces at the river Granicus. This demonstration convinced most of the cities of Asia Minor. They opened their gates to the victor without putting up a fight, and welcomed Alexander as their liberator from the oppressive and hated Persians. But Tyre refused to give in, and fought desperately for seven months to keep the Macedonians out. Alexander was infuriated and inflamed to the verge of insanity. He vowed to make an example of Tyre that would put the fear of whatever god he had—himself—into all who might oppose his divine will. When the city fell, Alexander gave a striking exhibition of his poor sportsmanship. He slaughtered all males capable of bearing arms. Then, borrowing the idea from the Achilles-Hector episode in Homer's *Iliad* (which he always carried with him and read on his campaigns), he tied the brave but vanquished defender of the city to his chariot and dragged him round the walls. To emphasize his point, he then razed Tyre to the ground.

Darius, King of Persia, with his main army of 600,000, was next on Alexander's schedule. To keep the record straight and award dishonor where it is overdue, it must be mentioned that this Darius was not the great one, but only Darius III, Codomannus, who had no greatness at all. Alexander and Darius collided head on in the pass of Issus. Darius lost the battle. The spoils of war included the family of Darius. The victor treated them considerately, perhaps because there were several royal and luscious females in the booty.

How were such incredible victories possible? Partly, but not wholly, by virtue of Alexander's skill as a general. Alexander's army was made up of mercenaries, mostly Macedonians. The prospect of rich loot kept them loyal and made them almost unbeatable; that, and a rigid discipline. They did not fight for slogans; they had none. They would have understood perfectly the Chinese soldier taken prisoner in Korea: when asked by an intelligence officer of the United Nations what he was fighting for, the prisoner replied quite simply, "Women and loot." The Medes and the Persians also had a rigid discipline, as attested by the avenues of crucifixes with their hanging cadavers through which the Grand Army of the Empire marched. But a stupidly administered discipline may be less effective than none. It was the Macedonian mercenaries who undid the Persian hordes by showing them how to fight and then deserting them. Why should they fight *for* the Persians when there was so much more to be gained by fighting *against* them? No reason at all. So they changed sides, but not for any loyalty to Alexander. Without the Macedonian phalanx, that hard core of tough fighters, to take the lead in battle under competent commanders, the helpless Persians milled about like sheep in a slaughter yard. The Macedonians butchered as many as they had time for before marching on to another kill.

Alexander was next attracted by the ancient and wealthy city of Damascus, which he occupied peaceably and robbed of all its money. Report says that he forwarded the very considerable sum of 600 gold talents to Aristotle to finance the Lyceum which Aristotle was setting up in competition with Plato's Academy. In view of Alexander's growing dislike for his former teacher at this time, the report may be untrue.

When he had recovered from his Homeric tantrum at Tyre, Alexander resumed his march. His next foray led through Palestine and took him to Egypt. The Egyptians saw in him a liberator from the tyrannical Persians, and welcomed him warmly and sincerely. This was a new experience for Alexander. He reciprocated by treating them decently and restoring their cherished customs, which the Persians had suppressed. While in Egypt he made one of his great contributions to the advancement of civilization, by founding (332 B.C.) the maritime City of Alexandria on the Mediterranean just west of one of the mouths of the Nile. Though he could not have foreseen what Alexandria was to become, he had taken the first step toward its intellectual supremacy. Being in a hurry to overtake Darius and subdue Persia, he did not tarry to build the metropolitan city of his dreams. In all, Alexander founded seventeen Alexandrias, of which only the one on the Nile and Kandahar (a corruption of *Alexander*) in the Punjab are remembered; the others were forgotten not long after their founder's timely death. Assured that the reforms he had introduced would make it comparatively easy for his more or less enlightened successors, if any, to govern the country efficiently for their own gain and glory, Alexander left Egypt. His objective was the total overthrow of Persia.

Near Arbela he caught up with Darius and his huge new army—some say as many as a million and a half men. The ill-trained horde was utterly defeated. Darius was swept along in the panicked rout. When his chariot horses bolted he made an undignified exit from the field of battle and from history. The great treasure cities of Babylon and Susa capitulated without even a token resistance. They actually welcomed the conqueror and practically begged him to plunder them.

5

Cleopatra's Gift

Though dead, Alexander was not yet through with Alexandria. While in Egypt, he had consulted an oracle in the Oasis of Amon and, partly to please the Egyptians and deify himself, had adopted the religion of Amon. He gave explicit orders that his body was to be buried in the Oasis. On his sudden death his sprawling empire quickly fell to pieces. His generals divided the wreckage among themselves. The African empire, including Egypt, fell to Ptolemy I, Soter. He had Alexander's cadaver sheathed in gold and transported south through Syria and Palestine—an unnecessary detour—so as to pass through Egypt on the way to the Oasis. The Oasis, not in Egyptian territory, lay in a neutral waste. Soter had good reason to covet Alexander's corpse. It was a palladium: whoever possessed it was assured of victory. Disregarding Alexander's orders—he was dead so mutiny was safe—Soter assembled all his forces and set out to meet the funeral procession, ostensibly to provide a worthy escort to the Oasis and to do his late commander whatever honor was due him. As his army greatly outnumbered the funeral cortege, Soter safely absconded with the corpse, to which he had no right. By-passing the Oasis, he made straight for Egypt. On crossing the border into Egypt, Soter began telling the truth, not that it made much difference, as nobody had believed him about the Oasis. He had already arranged for obsequies in Alexandria, including a stupendous procession with music to match and, appropriately enough, as much free wine as the thirstiest spectator could hold without vomiting.

There seemed to be no end to this greatest show on earth and in history. Samples of all the animals, including the military elephants and the unbelievable camelopards (giraffes), Alexander had seen on his travels were tastefully sandwiched in between marching cohorts of representative captives from Sparta to Babylon and all stations east. The procession took several hours to pass a given point. No Chicago gangster ever had so gorgeous a funeral, nor had Barnum and Bailey, the greatest showmen in American history, ever imagined so magnificent a circus. It seems a pity that Olympias had to miss the show, but she was unavoidably detained in the underworld. The sarcophagus, with Alexander's golden sheath exposed to view, was deposited in the civic center. There, till Roman times, it was gawked at by yokels and unintelligentsia alike. Finally the sheath and its sacred contents disappeared. Nobody seemed to know exactly or to care greatly what had become of either of them. A disreputable character was suspected of stealing both but was not caught. Considering the Moslem capture of Alexandria in A.D. 641, a believer in Greek superstitions might think that the palladium had somehow fallen into the custody of Mohammed or one of his capable generals. With his right arm tied behind his back, Alexander could have wiped the Moslem hosts from the face of the earth.

After Alexander, Ptolemy (Soter). For three centuries the Ptolemaic dynasty governed Egypt. Ptolemy I (Soter, 367?–

283 B.C.) founded the dynasty; Cleopatra (69–30 B.C.) saw its collapse under Roman robbery. Both Soter and Cleopatra fostered the magnificent Alexandrian Library. Cleopatra had nothing to do with the great Museum, of which more later; Soter saw that it was founded. It seems strange that Cleopatra, who could think with the outside of her head and had but little use for books, should have given the Library a desperately needed second start after an original Library was destroyed—inadvertently, no doubt—by Julius Caesar, the most notorious Roman of them all. But perhaps the honor of perpetuating the Library should not be awarded to Cleopatra but, ironically, to the stupidest oaf in history. This will appear when we reach Roman times and the last of the Ptolemies.

Alexander, as we saw, did not tarry to supervise the actual building of Alexandria. He left at least the first stages of that colossal job to Cleomenes, one of his generals, while he himself hastened after the elusive Darius. Cleomenes proved himself both competent and crooked. His dishonesty perhaps did not matter much. The building progressed almost too rapidly; and when Soter found time to look into things he detected several swindles, and promptly liquidated Cleomenes. The buildings for the Museum and the Library were completed about 300 B.C., only six years after they had been imagined. Even today, with modern machinery, this achievement would be noteworthy; with only slave labor it was impressive. The entire city was planned on a generous scale. According to Strabo (64 B.C.–A.D. 21, both dates doubtful), whose life partly overlapped Cleopatra's, two main avenues, each a hundred feet wide and lined with colonnades, divided the city into quarters. The buildings were flat-roofed, of stone, in blocks one by three miles in area. The civic center, given over to temples and the royal buildings, occupied a fourth of the entire site. There were two harbors, crowded with more shipping than any other port in the ancient world had. Alexandria dominated the eastern Mediterranean and the surrounding countries commercially, and led the world intellectually till the Roman conquest (first century B.C.) reduced Egypt to the status of a plundered Roman colony.

Much of all the wealth and splendor of Alexandria in its glory was due to the initial impulse imparted by Soter. His immediate personal contribution to Alexandria, however, was not in the material realm of stone and brick, but in the applied social sciences.

On taking over Alexander's African empire, Soter imported a swarm of Greek soldiers, practical engineers and civil servants, sycophants, and Athenian homosexuals. All this avaricious and wrangling mob he dumped into Alexandria to pacify, loot and govern Egypt and the adjacent territory. Alexandria already had a considerable Greek population when the unwelcome newcomers arrived. There was also a large segregated colony of Jews who, curiously enough, were called Hellenes by their cousins in Palestine. When her despotic turn came, Cleopatra emphasized her Greek dislike of Jews by rationing their lawful share of the available grain to a fourth of the acknowledged subsistence minimum. Though Queen of Egypt, Cleopatra was a Macedonian Greek of the same royal but now diluted stock as Alexander and Soter. The Egyptians were restricted to their own quarter lest the sight of these displaced persons should offend the sensitive and supercilious Greeks. They were merely "the natives," and as such were denied any opportunity to contribute to the greatness or prosperity of Alexandria. Yet in spite of all these disabilities the native Egyptians were assured at least of impartial justice by the earlier Ptolemies.

Soter's policy was to hold onto Egypt with his left hand while reaching out with his right to grab more. For fifty years it was war, war, war, from Cyrenaica to Cyprus and Palestine. Soter ventured some desperate gambles but nearly always won, either by brilliant generalship or, in the diplomatic sphere, astute chicanery. He was a thorough disbeliever in the comforting doctrine that a state can be governed and held together solely by pious platitudes, no matter how noble and lofty. Contrary to what those who profess to hate war might wish to believe, Soter's incessant wars were a good thing for Alexandria. The city prospered materially and culturally. How did this disconcerting miracle come about? Quite naturally, as will appear in a moment.

Was the idea for the Alexandrian Museum with its Library Soter's own? Or did he get it from someone else? It must be remembered that Soter was not merely a brilliant soldier but a highly intelligent man with more than a dabbler's interest in the science and literature of his time. He himself wrote the history of Alexander's campaigns, now unfortunately lost. He was a hardy and jovial sinner with a discriminating taste in what he considered the good things of life, including wine, women, and books. The first two were enjoyed by all orthodox Ptolemies, the books by only a few. Soter's personal collection of books is said to have been donated as a nucleus to the Library. On the whole, Soter has a reasonable claim to the honor of founding the Library.

But as usual in claims to historical distinction, there are possible "sources" other than the obvious ones. Even Aristotle has his partisans, on the basis of that memorable education Soter and Alexander escaped from when Papa Philip cut it short. Aristotle is said to have been the first to think of making a collection of books and founding a library. We know now that Assurbanipal, King of Assyria, beat him by about 300

115

years. May it not be possible that Soter heard of that great cuneiform library while he was in the East? Though the possibility is remote, Assurbanipal seems no farther off than Aristotle as a progenitor of the Library. However, all trails in the history of science must be followed, even if they sometimes lead only to mares' nests. Though Soter may be acquitted of Aristotle, Aristotle is still obligated to Soter. The Library became the storehouse and fountainhead of Aristotle's philosophy, whence it seeped into the European Middle Ages. This must give both Soter and Aristotle a big belly laugh, wherever they may be.

A more likely contender for Soter's idea is Demetrius Phalareus (345?–283 B.C.), an Athenian gentleman of great wealth, vast learning, and undeflatable oratory. When the Athenians could absorb no more of his gas, they kicked Demetrius out of his luxurious palace to wander at will. He trudged to Alexandria. Soter welcomed the outcast to his court and treated him royally. In return for this generosity, Demetrius gave Soter the idea of founding the Museum and the Library. Another tradition, about which there is no doubt, concerns Soter's son and successor, Philadelphus, who was to cram the Library with all the books he could buy or borrow and have transcribed into Greek. Soter asked Demetrius to which of his sons he should bequeath his kingdom. Though Philadelphus was not in direct line of succession, Demetrius named him, and Soter assented. It was a fortunate decision, except for the disinherited heirs, who grumbled. On taking office, Philadelphus stopped all murmuring, in the classical manner of Greek tragedy, by murdering two of his brothers, thereby silencing opposition and securing himself firmly on his throne. Demetrius

116

escaped the displeasure of the losers by fast running. But his wind was not what it had been in his oratorical prime. He was already something of a sot. As alcohol is said to be a cure for snakebite, Demetrius, in his philosophic muddlement, may have theorized that the remedy should work both ways, and snakebite be a cure for alcoholism. He tested his theory. It worked. Like Cleopatra, he used a hoodless cobra. He attained a posthumous fame with the school of oratory which he persuaded Soter to include in the curriculum of the Museum. Except as a wry joke, Soter could never have thought of a trick like that by himself.

Soter died in 283 B.C. at the age of eighty-four. Two years before his death he had abdicated in favor of his son Philadelphus (309?–246 B.C.), who thus became the second Ptolemy to rule Egypt. Philadelphus was twenty-three when he took his royal seat. He reigned from 285 to 246, a span of thirty-nine years. To start the new King of Egypt off in proper style Soter, a showman of genius, mounted a stupendous procession. This gorgeous spectacle was a living and moving catalogue of all the strange things Soter had seen during his long and active life, and a synopsis of his conquests. No doubt Philadelphus was sufficiently impressed by his father's farewell performance. Until he saw it, he might not have believed that the old man had all this in him.

Philadelphus was no soldier like Soter. He would fight when provoked by necessity, but he was a lover of peace and ease. The Museum and the Library were his passion, not all those ostriches and elephants in his father's procession that astonished and delighted the people he was to rule. He would pursue peace because war interfered with his sound love of

117

pleasure. His taste in mistresses, both Greek and Egyptian, was catholic even for a Ptolemy, and his collection set a record. War could only distract him from the really worth-while things in life. He was himself a patron of the arts and humanities, so he knew the value and uses of books. As he kept Egypt out of costly wars, the country prospered and the people left him to cultivate his hobbies undistracted. The only shadow on his happiness was a morbid fear of death. The certainty that he must soon join his murdered brothers frequently reduced him to tears. The sight of young boys wrestling in the sand was more than he could bear; someday he would be dead, while they would be living, vigorous men. Egypt, with its mummies and its uncounted monuments and memorials to the dead, was a vast necropolis in which he was compelled against his will to be aware always of what he dreaded and could not avert. Like his father, he was remarkably free of superstition. He did not supplicate any gods to deliver him, because he believed in none. Having only himself to call upon, he grasped whatever pleasure he could find and got on with his plans for the Museum and the Library.

Though Philadelphus probably never heard of Assurbanipal's great library, his own methods in assembling the Alexandrian Library might have been copied from those of the scholarly and peace-loving Assyrian. Actually there were two Alexandrian libraries, one in the Bruchium quarter on the waterfront; the other, later, was in the Serapeium (temple of Serapis) in the Rhacotis quarter. Both came to disastrous ends, as will appear in the proper places. Philadelphus was responsible for the first, lodged in the Museum. The Alexandrians had one inestimable advantage over the Assyrians: papyrus. This natural "paper" made possible the compact and easily legible "books" (properly, rolls) which preserved and transmitted the lore of their epoch. The Egyptian invention of papyrus as writing material was as long a step beyond the clay tables of the Babylonians and Assyrians as was the invention of printing beyond handwritten manuscripts.

Sparing neither wealth nor brains to acquire the best obtainable, Philadelphus' agents scoured Europe and Asia Minor for books. Tourists and others bringing books into Egypt were promptly relieved of them for the benefit of the Library. But they were not sent bookless away. The diligent scribes on the Library staff presented them with free copies of their confiscated property. This generosity may account for some of the varied "readings" of passages in the ancient classics that for centuries have provided scholars with interminable disputes and respectable livings. On another level is the hope that a lost work by Euclid or Archimedes may turn up in the rubbish heaps or overlooked repositories of the past. One such did, a revealing treatise by Archimedes on his "method," as late as 1906 in Constantinople.

The early Ptolemies poured books into the Library at such a rate that over 600,000 were accumulated in the first forty years. There seems to be no way of expressing the figure in equivalents of our printed books. No catalogue of what the Library contained survives. That it was the most extensive collection of the ancient world is generally accepted. Works in languages other than Greek were translated. Unlike the narrow and self-satisfied Athenian Greeks, the liberal and modest Alexandrians welcomed the sciences and learning of all peoples who had any. Consequently the Library became an unsurpassed treasure house of the science and scholarship of the ancient world, with the possible exceptions of some of the countries Alexander visited on his campaigns. If only the Library could have survived Roman indifference and Christian intolerance, the Dark Ages need not have happened so soon, if ever. But Western civilization broke down under the collapse of a decadent empire and the bigoted hostility of a proletarian religion.

The Museum was under the direct patronage of Philadelphus. He induced some of the prominent Greek scholars and scientists, including mathematicians, to migrate from Athens to Alexandria. Promises of favourable working conditions were not only made but, what was remarkable at the time or indeed at any time, kept. Those who accepted Philadelphus' invitation were accommodated in the Museum, which had some of the features of a modern university but was far more than any university, ancient or modern. Living quarters and dining halls were provided for the permanent staff and visiting intellectuals. Staff members were free to carry on their researches in any way they chose, and might lecture or not as they pleased. The dominating policy of the Museum was the advancement of science and learning. Nothing else counted. There were four "faculties": literary studies, including poetry and oratory, medicine, astronomy and mathematics. Soter, while planning the Museum, is said to have consulted Demetrius on what studies should be encouraged. The inclusion of oratory certainly suggests Demertius, not Soter, whose eloquence was of the practical kind that got results at the point of a sword if necessary. It seems as if Soter must have given into his adviser's oratory when he could take no more of it. Actually, the faculty of poetry and oratory was the weakest of the four. This may not have been the fault of the members; the great age of Greek literature was already in the past when the Museum opened, and all the bewildered scholars could contribute was criticism, appreciation of the masterpieces of the past, and footnotes. However, the faculty of oratory did train some pretty voluble and orotund public speakers. The most durable of the four faculties, to judge by what has survived, were mathematics and astronomy. The closest modern equivalent of the Museum as a whole was the Kaiser Wilhelm Gesellschaft in Berlin, financed (at the Kaiser's "request") by rich German industrialists and other men of wealth. Two World Wars effectively finished off the German institute; the Museum was destroyed in only one very minor war.

There is more of the story of the Museum to come, but this is an opportune chance to cast up the account so far and strike a trial balance.

The native Egyptians under the Ptolemies were submerged in virtual slavery. The state was in actual or potential possession of everything, including the bodies of the peasantry. The country crawled with tax collectors extorting all they could get their hands on, leaving barely enough of the crops to keep the serfs working. The economy was an echo of ancient Egypt under Cheops (if we accept Herodotus' account), and a preview of Fermat's seventeenth-century France. The government was corrupt to the point of complete rottenness, and the people suffered accordingly. To escape slavery and the perservering brutality of the tax collectors, backed by the military, the producers fled to the swamps of the Nile and lived on what they could grub from the muck, or starved to death. Had the serfs been able to get arms, there would have been a revolution on the grand scale, as in eighteenth-century France or twentieth-century Russia. As they had nothing but their hands and crude, ineffectual farm implements to fight with, any sporadic uprisings were quickly put down by total slaughter. How does this picture square with that of even-handed justice for all human beings in the land of Egypt under the Ptolemies? Perfectly: slaves were only subhuman and so had no human rights or privileges. The dignity of man and the sanctity of human life were not even idle dreams as yet. Such was Egypt

121

under the Ptolemies who lived in almost Persian luxury and gave civilization the Museum and the Library at no cost to themselves. So much for one side of the account. Proponents of this side say the cost of the mathematics and the rest that came out of Alexandria was inconsiderable when weighed against the benefits that accrued to future centuries of civilization. The other side, represented by the oppressed, the starved, the tortured and the massacred says nothing, because it was exterminated before it could say anything.

About halfway through its three centuries of rule the Ptolemaic dynasty degenerated in extravagant luxury and enervating debauchery. Toward the end of this decline, the still-virile Romans came clumping into Egypt and took over Alexandria. By this time Alexandria, among its other titles to fame, was renowned as "the granary of Rome." It might have been to the ultimate advantage of civilization if Alexandria had let Rome starve.

As Cleopatra was the last of the Ptolemies and as through her, if only rather indirectly, the great Library was renovated, the concluding years of her life merit a summary notice. The picture of her last days in Shakespeare's *Antony and Cleopatra* portrays her substantially as the chroniclers have done. But Shakespeare adds something that is in no fusty chronicle.

A full account, for which there is not space here, would do justice to the Roman occupation of Egypt, which sometimes resembled a drunken brawl between brutalized soldiers in a bawdy house. The struggle between the Egyptians and the Roman invaders of Egypt for the possession of Alexandria was sharp and bloody. The Egyptians lost. Only three of the Romans, Julius Caesar (102–44 B.C.), Mark Antony (83?–30 B.C.) and Octavianus, later "the Great Augustus" (63 B.C.–A.D. 14), need concern us here. Of the Egyptians, only the

Queen, Cleopatra, is important to us. Her charm and beauty have been overplayed at the expense of her physical courage and her brains; she was as brave as any man and far more intelligent than all but a few. Small wonder the loutish Romans found her irresistible. She was ruthless and unscrupulous when she had to be to preserve her kingdom. She also had a streak of Macedonian cruelty in her that shocked even Antony on occasion. A messenger bringing bad news either had his throat cut by a flick of her own jeweled sword or was elaborately tortured to death. To make sure that her food was not poisoned, a slave had to eat some of it. That of course was standard court procedure. Cleopatra sharpened her appetite by watching the slave's death agonies if the food really was poisoned; if not, she gave him a telling slash of the sword for having deprived her of an anticipated pleasure. But having seen Olympias and her son Alexander in action, we need not be finicky about Cleopatra when she was merely following the family tradition.

When the Romans arrived, Cleopatra deployed all her forces to preserve her kingdom. She took them on single-handed. Before a year had passed, she capped Caesar's famous boast, "I came, I saw, I conquered" (*Veni, vidi, vici*), with one of her own, but not in public, "He came, he saw, I conquered." In fact her first notable conquest was Julius Caesar, who had got his start toward the top by submissive pederasty. Young Julius literally had begun at the bottom, and had risen like a rocket to love, fame and glory. At the age of eighteen, a handsome fledgling Roman legionnaire, he was willingly seduced by King Nicomedes of Bithynia. Julius Caesar was no green innocent; he knew what he was doing. By "sleeping with" the robust and virile king, he could get himself up a step or two on the ladder to power and fame. His gamble paid off. Nicomedes, it is said, was disappointed in his chosen love. Not so Cleopatra; she could not get enough of him, nor he of her. Caesar's weakness, after he became a sexually normal man, was beautiful women. And now he had come to conquer Egypt and, incidentally, its Queen, Cleopatra.

She reduced him in short order to what, for him, was imbecility. He lived with her over a year. Always accommodating, Cleopatra bore him a son, Caesarion. By the Roman law of succession through the male heir, Caesarion should have been Caesar's legitimate successor. Egyptian law prudently legitimized the female line; there might be some doubt about a child's father but none about its mother. Lacking a daughter by Cleopatra to succeed her as Queen of Egypt, Caesar settled for the next most likely heir to the throne of Cleopatra, her sister Arsinoë IV. When he finally tore himself loose from Cleopatra, Caesar took Arsinoë back to Rome with him, draped her in chains, and gave her the place of dishonor directly behind his chariot in the triumphal procession celebrating his subjugation of Egypt. This was one of his few shortsighted acts. By all the rules of the royal game he should have put Arsinoë to the sword to cut off possible disputes over Egyptian succession. But he spared her to breed possible trouble.

During Caesar's dalliance in Egypt, Mark Antony acted as his delegate in Rome. When Caesar returned, Antony prudently left. Cleopatra thought it would be nice to pay Caesar a visit in Rome. She took her younger brother, who was also her legal husband, with her. When Caesar was assassinated, Cleopatra found herself unwelcome and returned hastily to Egypt with Caesar's son Caesarion, the lawful heir to the throne of Rome. To shut off probable disputes about the succession, she had her boy husband-brother murdered when he reached the age of fifteen.

Antony in person enters here. He was fighting the faction led by the assassins of Caesar. Distrusting Cleopatra, he sent her a curt order to appear before him at Tarsus in Cilicia, and talk herself out of any complicity in the conspiracy that had eliminated Caesar, if she could. She not only could but did. She arrayed herself in the finest of her finery, enhanced her skin with exciting perfumes and her very best jewels. Then she boarded her royal barge, sailed with the Egyptian fleet, and descended on the hapless Antony. Foolhardily he received her in private. She was twenty-eight at the time, at the peak of her physical and persuasive powers. The barge was as described by Shakespeare. The honest and stupid Roman soldier Antony had seen nothing like it, or her, in all his rugged life. Antony was a pushover for Cleopatra. If there is any moral in this or in what happened to Caesar, it is that generals on campaign should keep away from beautiful and fascinating women. Cleopatra had made a limited fool of Caesar. With that trial run behind her, she would now make a complete fool of Antony.

Tarsus was no place to stage a torrid flirtation, so the proceedings adjourned to Cleopatra's place in Alexandria. There she had her womanly way with Antony. He was already worn down by drink and debauchery, but Cleopatra found him worth cultivating, for political reasons if for no other. In his blind and frequently drunken infatuation he promised her anything she might ask. Her first request was that Antony should have Cleopatra's sister Arsinoë killed. Antony was momentarily shocked, but he could not go back on his promise. So the helpless Arsinoë was duly murdered in—of all places—the temple of Diana in Ephesus where she had sought sanctuary. "Great is Diana of the Ephesians!"

Another of Cleopatra's requests was more civilized (according to our professed morals) and of vastly greater significance for the future of mathematics and learning. The Museum with its Library in the Bruchium quarter was destroyed in 47 B.C., when Caesar fired the Alexandrian fleet in the harbor. Some say this mishap was an accident; others say that the Roman soldiers deliberately started the fire. As the truth is unlikely ever to be known, we may believe what we please. If any good could come out of such a disaster, some did. It must be remembered that Cleopatra was of the Ptolemies and shared their respect for science and learning. Without intellectual supremacy, Alexandria was not much better than another Rome. Could nothing be done to restore the Library, she asked Antony. He offered a practical suggestion. The city of Pergamum in Asia Minor had hoped to rival, if not to surpass, Alexandria as a cultural metropolis, and had accumulated a splendid library of some 200,000 books. "Get it for me!" Cleopatra demanded. Antony could not refuse. So Pergamum lost its library and its hope of ever being anything but just another mercantile town. Cleopatra housed the books in the temple of Serapis (the Serapeium), and Alexandria again had the leading library in the world. Before very long the nucleus of 200,000 stolen books had swelled to about 700,000. We shall see presently what happened to the Library of the Serapeium. For a moment we must follow Cleopatra and her devoted Antony to their mutual end.

To impress the Romans and please her lover, Cleopatra put on one incredible banquet after another for Antony and his generals. She presented the gold plate and the gold furniture on which they reclined to her astonished and boorishly greedy guests. They had never seen or imagined such luxury and

In 642, after a siege of fourteen months, Alexandria capitulated to the Moslems. What was left of the Library—probably very little if Orosius reported accurately—was destroyed. The decision of what to do with the books had been put to the Caliph Omar by the victorious Moslem general. What immediately follows makes sense only if a considerable number of books were found. "As for the books you mention," Omar replied to the general, "if they contain what conforms to the Book of God [the Koran], the Book is sufficient without them; if they contain what is contrary to the Book of God, there is no need for them. So give orders for their destruction." The Moslems thereupon began stoking the books into the furnaces of the numerous public baths in Alexandria. It took several months to burn up all the books.

Whether true or false, this story preserves or invents the perfect gem of Aristotelian logic attributed to Omar. Soter would have appreciated it as a tribute to his old teacher. So it may fittingly close our account of the Alexandrian Library from Soter to Cleopatra. And before taking sides in the historical controversy provoked by the story, we may well remember that a thoroughly reliable historian would have to be omniscient.

I pass on to the arithmeticians associated in one way or another with Alexandria.

6

From Euclid to Hypatia

For easy reference I list here the names and dates of the Alexandrian arithmeticians from Euclid to Hypatia whom we must consider on our way from Babylon to France.

Euclid, born about 330 B.C., died 275 B.C.
Archimedes, born about 287 B.C., died 212 B.C.
Eratosthenes, 275–194 B.C.
Apollonius, born 260 B.C., died about 200 B.C.
Nicomachus, born A.D. 50, died about 110.
Diophantus, first (?) century A.D. (See Chapter 7, Part 1, here.)
Iamblichus, flourished about A.D. 310.
Hypatia, murdered A.D. 415.

For readers acquainted with the history of Greek mathematics this list may contain two surprises, the inclusion of Apollonius and the disputed date of Diophantus. Apollonius is not usually mentioned in connection with the theory of numbers; his only contact with the subject is the satirical joke Archimedes is supposed to have played on him. Diophantus until the late 1930s was placed as far forward as the second half of the third century A.D. Modern historical research has *temporarily* moved him back to the first century. The major figures here are Euclid and Diophantus. Again I emphasize that we are primarily interested only in what concerns, even if occasionally somewhat remotely, the theory of numbers, especially those items having a bearing on the work of Fermat and his contemporaries in the seventeenth century. Improvements in the art of computation, however important for astronomy and interesting in themselves, are irrelevant for our purpose. It is the theory that counts here. Likewise for those contributions of the men discussed here to department of mathematics other than the theory of numbers—such as geometry and applied mathematics. Consequently what is described is a measure only of competence as an arithmetician, and is not intended to be indicative of the full stature of a man as a mathematician. Archimedes, for instance, made but a single rather minor contribution to the theory of numbers, while the followers of Nicomachus from their "bad eminence" in the Middle Ages surveyed many. I shall take the eight listed in their chronological order. Note that they span 845 years, and recall that modern mathematics originated only in the early seventeenth century, say about 300 years ago.

Euclid as a man and personality is all but unknown to us. Lacking facts, the Moslems and others invented fables to give the shadowy geometer some human substance. A fairly credible specimen must suffice. When Soter once asked Euclid if his proofs in geometry could not be shortened and made smoother, Euclid replied, "There is no royal road to geometry." One difficulty with this story is that it has been told of others. The few alleged facts in Euclid's life may also be inventions, for example the tradition that he studied mathematics in Athens under the Platonists. If he did, he, fortunately for the theory of numbers, did not transmit any of their numerology. In Euclid's severely mathematical treatment of numbers there is not a trace of Plato's fantastic metaphysical arithmetic. And much as we should like to credit the character sketch of Euclid by Pappus of Alexandria (born late third century A.D.), we note that it dates 500 or more years after Euclid's death. Euclid, according to Pappus, was scrupulously fair in giving credit where due to his predecessors and contemporaries, and was kind and helpful to serious students. Pappus, however, is suspected of using a fictitious Euclid to smirch the legendary Apollonius, whom he disliked, and who had none of the agreeable qualities he attributes to Euclid. One important detail of Euclid's life is not disputed: he was among the first to accept Soter's invitation to join the Museum staff at Alexandria, where he lived out his extraordinarily productive career. How one man in the course of a normal lifetime could accomplish all that Euclid did is the eighth wonder of the ancient world. If a ninth wonder is naturally suggested, it is Archimedes.

Euclid composed at least ten works on the pure mathematics of his day—geometry, arithmetica, "phenomena," optics, music. Reasonably complete texts of five of the ten survive. Only a small part of his masterpiece, his *Elements*, complete, concerns us here. In the *Elements* he collected, systematized and amplified much of the pure mathematics of his age and arranged it in logical order. It was the first great, even if far from satisfactory, essay in strict deductive reasoning. To dispose here of the inevitable and just criticism, I must add that numerous flaws and fallacies were detected by modern mathematicians in Euclid's supposedly rigorous chain of deductions.[1]

---

[1] In the early 1900s, when the teachers of mathematics in the English secondary schools were fighting to get rid of Euclid's geometry as a long-outdated textbook, they enlisted the help of Bertrand Russell, already an established logician and

These have been set right where it is worth while.[2] The main thing, and Euclid's personal contribution, has not been superseded. That was the attempted application of the axiomatic method to a large and heterogeneous mass of mathematics. Others before Euclid, notably the Pythagoreans and Plato's disciple Eudoxus (408–355 B.C.), were familiar with the method and had applied it fairly successfully, but Euclid's *Elements* was the first extensive application. The strictures on lack of rigor do not apply to the relatively few items concerned with the theory of numbers. They, at least, are mostly sound and without blemish. They also are basic for the theory of numbers as it exists today.

A source of difficulty in reading the *Elements* is the apparent lack of motivation. Euclid never says where he is going, nor even why he would wish to go anywhere. From the first definition to the last of the 465 propositions it is all strictly business with no side remarks. We may take what he offers or leave it as we please. Some commentators assert that his goal was the construction of the five so-called "Platonic bodies" —the five regular solids of 4, 6, 8, 12 and 20 sides investigated

---

mathematician, who could speak with authority. He wrote in part: "The claim [that 'Euclid's logical excellence is transcendent'] vanishes on a close inspection. His definitions do not always define, his axioms are not always indemonstrable, his demonstrations require many axioms of which he is quite unconscious. A valid proof retains its demonstrative force when no figure is drawn, but very many of Euclid's earlier proofs fail before this test." Russell then proceeds to the detailed demolition of Euclid's statements and fallacious "proofs" of the first seven propositions in Book 1 of his *Elements*. Russell concludes, "Many more general criticisms might be passed on Euclid's methods, and on his conception of Geometry; but the above definite fallacies seem sufficient to show that the value of his work as a masterpiece of logic has been very grossly exaggerated."—*Mathematical Gazette*, II (1902), 165–167.

[2] Frequently at the cost of discarding large sections and making a fresh start. Book 5, "the crown of Greek mathematics," is beyond repair. See concluding chapter here.

by the Pythagoreans—which he gives in Book 13, where he shows how they may be inscribed in a sphere. He proves that these five are the only regular solids possible. If such were indeed his goal, he might have chosen a shorter route and satisfied Soter with a good first approximation to a royal road. Many of his theorems are not only superfluous but, what is worse, dull, as must have been noticed by thousands of hapless schoolboys being led or cuffed and flogged through Euclid. For 2200 years scraps of the geometrical books of the *Elements* were the gospel of school geometry, and "Euclid" became a synonym for elementary plane geometry.

Before the invention of printing, numerous manuscript copies of the *Elements* were available for study. The first printed edition was a Latin translation from the Arabic in 1482 —ten years before Columbus discovered America. It was followed down to our own times by over a thousand more. We may safely wager that there will not be a second thousand editions, bombs or no bombs.

The *Elements* consists of thirteen books, of which numbers 1, 3, 4, 6, 11, 12, 13 expound elementary plane and solid geometry. Books 1, 3, 4, 6, and a few propositions from 11, 12 contain the substance of the standard and antiquated school course in geometry today. The theme of Book 2 is a treatment of certain simple algebraic matters, such as $(a + b)^2 = a^2 + b^2 + 2ab$, expressed by Euclid in an awkward and repellent geometrical language. These are now derived in a civilized course by a few lines of easy algebra. Book 5 contains the Eudoxian theory of proportion, "the crown of Greek mathematics." Euclid needs some of it for his Book 6. As it is much too hard for immature students, it was early dropped from the texts in elementary geometry. Its essential content in modernized form is now part of the difficult theory of the "real number system," which is excluded from the theory of numbers in our current sense of the term. Similar remarks apply to Book 10, which connoisseurs of crabbed Greek mathematics prize as the gem of the whole work. Only the hardiest and most obstinate scholars can get through it in the original today. (Fermat despised it as mathematics.) Book 8 deals with numbers (or "magnitudes") in geometrical progression. Its essential substance is now simply developed in the usual short chapter on geometrical progressions in school algebras. The one detail we need note is the "L.C.M.," to be defined presently in connection with the "G.C.D."

The items of interest here are a few basic theorems on numbers in Books 7, 9. Numbers are represented by segments of straight lines. The segments are designated by single letters, so the treatment is general. The proofs, given in geometrical language, are now much more simply presented by means of the algebraic symbolism which Euclid lacked. It is long since his proofs were read seriously by any but specialists in the history of Greek mathematics. The theorems, however, are as penetrating and vital today as when Euclid stated and proved them, and indeed are indispensable in any development of the current theory of numbers. If some of those noted presently seem self-evident or even trivial to the reader, he should try to prove them without consulting a book. He may revise his estimate.

It will be well here to refer to the definitions in Chapter 1 of divisor, prime, composite, aliquot parts, factor, and perfect number. As always, *number*, without qualification, means positive integer. I must now recall another definition, and explain it.

The *greatest common divisor* (G.C.D.) of two or more numbers is the largest number that divides each of them.

137

Thus the G.C.D. of 6 and 21 is 3; the G.C.D. of 22, 55, 132, is 11. For, all the divisors of 6 are 1, 2, 3, 6; all those of 21 are 1, 3, 7, 21, and the largest number appearing in both sets of divisors is 3. All the divisors of 22 are 1, 2, 11, 22; all those of 55 are 1, 5, 11, 55; all those of 132 are 1, 2, 3, 4, 6, 11, 12, 22, 33, 44, 66, 132, and the largest number in all three sets is 11. These G.C.D.s were obtained here essentially by trial. Unless we had succeeded in finding the divisors of the numbers concerned we could not have proceeded. But, as Euclid showed, the difficulty can be obviated.

Suppose we had to find the G.C.D. of two fairly large numbers, say 72829807 and 75185353. We might spend several hours trying to find all the divisors of these numbers, especially if they should turn out to be primes, when it might take days. Without first finding the divisors of the numbers concerned, Euclid in the first three propositions of Book 7 of his *Elements* gave a straightforward process for getting the G.C.D. of any 2 or 3 numbers (and therefore of 4 or 5 or 6, or ..., by repetitions of the process in an obvious way). This seems like a mathematical miracle. The process is called the *Euclidean algorithm*. Using only repeated divisions, multiplications, and subtractions in a prescribed order, the algorithm is nontentative and no more complicated than long division. Any moderately quick twelve-year-old could find the G.C.D. (it is 8353, incidentally a prime) of the two numbers above in a few minutes. The school algebras used to contain Euclid's algorithm, and even more arithmetics at the turn of the present century explained it. It was dropped from the elementary texts because there seldom is any use for it in applications this side of the theory of numbers and modern algebra, where it is indispensable.

Closely allied with the G.C.D. is the L.C.M. (*least common multiple*) of two numbers. It is the least number divisible by each of the two numbers. Euclid discussed—rather laboriously —the L.C.M. in his Book 8. Unless I have overlooked it in all the involved verbiage, he missed the interesting theorem that the product of the G.C.D. and the L.C.M. of two numbers is equal to the product of the two numbers. Example: the G.C.D. of 12 and 18 is 6; the L.C.M. is 36; $6 \times 36 = 12 \times 18$. By using this theorem we can calculate the L.C.M. of two numbers a, b without having to find the divisors of either: if l is the L.C.M., and g is the G.C.D., $l \times g = a \times b$, $l = (a \times b)/g$, and g is founded by Euclid's algorithm.

We now need two more definitions, the last for some time. If the G.C.D. of two numbers is l, the numbers are called *relatively prime* (to each other), or *coprime*.[3] For example, 30 and 77 are coprime. Each of two coprime numbers, as implied above, is said to be *prime to* the other. Thus 30 is prime to 77, and 77 is prime to 30.

The following theorems (numbers 24, 27, 28, 29, 30, 31, 32) from Book 7 are the basis of the theory of arithmetical divisibility. Without these there would be no theory of numbers as we know it. They are often given in the first chapter of texts on the theory of numbers.

Theorem 24   *If each of two numbers is prime to any number, their product is prime to that number.*

Example: Each of 21, 69 is prime to 100, as is their product 21×69 or 1449.

Theorem 27   *If two numbers are coprime, so are all their successive powers.*

---

[3]This convenient term, for several reasons, is preferable to the other. It is increasingly used in American writing.

Example: 12 and 35 are coprime; so also are 12^2 ($= 144$) and 35^2 ($= 1225$), 12^3 ($= 1728$) and 35^3 ($= 42875$), and so on.

Theorem 28   *If two numbers are coprime, their sum is co-prime with each of them; and if the sum is prime to either, the two numbers are coprime.*

Example: 22 ($= 2 \times 11$) and 69 ($= 3 \times 23$) are coprime; their sum 91 ($= 7 \times 13$) is coprime with each of 22, 69.

Theorem 29   *Any prime numbers is prime to any number which it does not divide.*

Example: The prime number 7 does not divide 15, and 7 is prime to 15.

Theorem 30   *If the product of two numbers is divisible by a prime, then [at least] one of the two numbers is divisible by that prime.*

Example: 11 divides 330, which is the product of 15 and 22; 11 divides 22, but not 15. Again, 7 divides 294, which is the product of the two numbers (among others) 14 and 21, each of which is divisible by 7.

This is one of the most useful theorems on divisibility. The proof follows from the definitions and theorems already given. The simplicity and "obviousness" of the theorems are deceptive. Failure to note all the implications of the proof was partly responsible in the nineteenth century for fallacious attempts by first-rate mathematicians to prove Fermat's Last Theorem. More will be said about this in the concluding chapter. For the moment the reader may like to prove (31) with only what Euclid knew and without any algebraic symbolism.[4]

---

[4] If you can do this entirely on your own without previous knowledge, you may give yourself an A. It isn't *always* true beyond the *rational* integers.

Theorem 31   *Any composite number is divisible by some prime*. Otherwise stated, a number other than 1, which is not a prime, has a prime divisor.

This may seem superfluous after the mere definitions of composite and prime, but it is not.

Examples: Suppose we know by some means that 1007273 is composite (not prime).

Then (31) assures us that this number has a *prime* divisor, even though we may be unable to find it. Actually the smallest prime divisor here is 101, which could be found by testing the odd primes $3, 5, 7, 11, \ldots, 101$ less than the square root of 1007273, as possible divisors. But suppose an exhaustive test were beyond our powers, even with calculating machines, as might be so for a number of the form P, where

$$P = 1 + 2 \times 3 \times 5 \times 7 \times 11 \times \cdots \times p,$$

and $2, 3, 5, 7, 11, \ldots, p$ are all the primes not exceeding the prime p. We might spend months or years testing for a prime divisor of P if p were only moderately large, say 9973. But P must be either prime or composite, and (31) tells us that if P is composite it must be divisible by *some* prime, even if we cannot find the prime in a long lifetime of testing. The proved *existence* of a prime divisor is sufficient for a decisive step in one of Euclid's most celebrated and finest theorems (20 in Book 9, stated and proved presently).

Theorem 32   *Any number either is prime or is divisible by some prime*.

Finally, in Book 7, we note Definition 22 where, for the first time on record, a perfect number is *defined* as a number which is equal to the sum of its aliquot parts. The Pythagoreans, as we saw, are held responsible for much numerology about what

141

they called "perfect numbers," but these were not perfect in Euclid's sense, which is ours today. For example, 10 for the Pythagoreans was perfect because of the mystical and cosmological properties ascribed to the equality $10 = 1 + 2 + 3 + 4$. According to tradition, they had observed that 6 and 28 are perfect in Euclid's sense; but these facts were merely the starting point for a flight into the blue of celestial numerology. We shall see shortly what Euclid got out of his unmystical definition.

The arithmetical propositions in Book 8, of but little intrinsic appeal, have long been superceded, at least in treatment, and are antiquated. They are of no interest to us here. But in Book 9 theorems 14, 20 are of capital importance for the theory of numbers, and theorem 36, the last in the book, is the earliest recorded significant fact concerning perfect numbers.

Theorem 14   *Every composite number is a product of primes.*
    Examples: $105 = 3 \times 5 \times 7$; $7272 = 2 \times 2 \times 2 \times 3 \times 3 \times 101$. In any such decomposition into primes, two products differing only in the arrangement of the primes are not considered distinct; for example $3 \times 5 \times 7$ and $5 \times 3 \times 7$.

Theorem 20   *There are an infinity of primes.*

Theorem 36   *If $2^p - 1$ is prime, $2^{p-1}(2^p - 1)$ is a perfect number.*
    Examples: It is readily seen that if $2^p - 1$ is prime, p must be prime. For $p = 2, 3, 5, 7$, $2^p - 1$ is 3, 7, 31, 127, each prime, and 2^{p-1} is 2, 4, 16, 64. So, according to (36), 2×3, 4×7, 16×31, 64×127, or 6, 28, 496, 8128 are perfect, as we noted before. Unfortunately (or fortunately?) it is not true that $2^p - 1$, where p is prime, is always prime; for example, $p = 11$,

and $2^{11} - 1 = 2047 = 23 \times 89$. We shall meet $2^p - 1$ again when we come to Mersenne in the seventeenth century—about 2000 years after Euclid. Euler in the eighteenth century proved that every *even* perfect number is of Euclid's form. We have noted (Chapter 1) the unsolved problem of *odd* perfect numbers. For an exercise, the reader may like to prove that an even perfect number must end in 6 or 8.

As the proof of (20) is simple and a good exercise in elementary (and outdated) classical logic, I give an old-fashioned version of it in detail, assuming that we have in mind Euclid's basic theorems on divisibility. The number P in the examples following (31) leaves the remainder 1 when divided by each of the primes $2, 3, 5, 7, 11, \ldots, p$, so none of these is a divisor of P. Suppose now that there are only a finite number of primes. Then there must be a largest prime. Call this prime p. By (32), P is either prime or is divisible by some prime. Clearly P is larger than p, because it is 1 plus some multiple of p. So if P is prime it is a prime larger than the assumed largest prime. This is a contradiction. If P is composite by (31) it must be divisible by some prime. But this prime, as we noted in the examples to (31), cannot be p or any one of the primes smaller than p. So it must be larger than p, again a contradiction. Thus the assumption that there is largest prime is untenable. So there is no largest prime, and this is equivalent to (20). As numerical examples of the two possibilities in the proof, P is prime for $p = 2, 3, 5, 7, 11$, so at the successive stages there are the new primes 3, 7, 31, 211, 2311; for $p = 13$, P is composite, $1 + 2 \times 3 \times 5 \times 7 \times 11 \times 13 = 30031 = 59 \times 509$, and we have the two new primes 59 and 509.

The proof of (36) goes as follows. *All* the divisors of $2^{p-1}(2^p - 1)$ are $1, 2, 2^2, \ldots, 2^{p-1}$ together with each of these

143

multiplied by the prime $2^p - 1$. The sum of all these is

$$\left(1 + 2 + 2^2 + \cdots + 2^{p-1}\right) + \left(1 + 2 + 2^2 + \cdots + 2^{p-1}\right)\left(2^p - 1\right).$$

and it is an easy exercise to show that this is equal to $2^p(2^p - 1)$. The sum of all the divisors includes the number $2^{p-1}(2^p - 1)$ itself, which must be subtracted from the sum of *all* to get the sum of the aliquot parts:

$$2^p\left(2^p - 1\right) - 2^{p-1}\left(2^p - 1\right)$$

which reduces to $2^{p-1}(2^p - 1)$. That is, $2^{p-1}(2^p - 1)$ is equal to the sum of its aliquot parts, and so is perfect.

Before leaving Euclid, we must look at (14) in Book 9, stated above. The proof, which I shall not reproduce, follows readily from the theorems given. From (14), by an argument which was beyond Euclid's resources, it is easily proved that *every composite number can be expressed as a product of primes in one way, and only in one way, if two arrangements of the same prime factors are not considered distinct*. This is usually called *the fundamental theorem of arithmetic*. Note that if 1 has not been excluded, by definition, from the primes, the fundamental theorem would not exist, as any number of 1's could be inserted as factors in the product.[5]

---

[5]The first explicit statement and proof of the fundamental theorem were by Gauss in his masterpiece, the *Disquisitiones Arithmeticae*, 1801, sec. 16. Before Gauss, the theorem had been tacitly assumed or at best vaguely described, as Gauss in effect remarks. T. L. Heath, in his critical and exhaustive edition of Euclid's *Elements*, incorrectly ascribed (I, 403) the theorem to Euclid. Anyone interested in the reasons why Euclid could not possibly have stated, much less proved, the theorem will find them in R. Fueter, *Synthetische Zahlentheorie*, 1921, p. 21. Since 1921 proofs by mathematical induction have been published. The earliest was devised in 1912 by E. Zermelo and printed in 1934. But mathematical induction was not invented till many centuries after Euclid. It seems rather remarkable that it was not, as it and Euclid's proof of (20) are exercises in one kind of logical tactics, which includes also Fermat's powerful method of infinite descent

Euclid was followed shortly by Archimedes, the most penetrating scientific, mathematical, and engineering mind of antiquity and one of those always included in any list of the three
greatest mathematicians in history. Except for a stay as a young
man in Alexandria, where he is said to have studied under the
successors of Euclid, Archimedes spent his life in Syracuse, the
capital of Sicily. He was a close friend of the royal family of
Syracuse and probably was related to it. This may account for
his thoroughly aristocratic temperament. If one of his scientific
or engineering inventions was of practical use, he despised it.
In pure mathematics, some parts of his work were not surpassed till the seventeenth century, when the integral calculus,
which he had anticipated in certain details, was developed by
Newton and Leibniz. The discoveries which he himself prized
above his others concerned the mensuration of the sphere, in
which he gave the first rigorous proofs (*for their age*) of the
now familiar expressions for the surface and the volume of any
sphere. These cost him a prolonged effort. Today they are
ten-minute exercises in the first course in the integral calculus.
We do progress in some things.

Archimedes thought so highly of what he considered his
masterpiece that he asked that his tomb bear the figure of a
sphere inscribed in a cylinder, to recall his classic theorems
that the volume of a sphere is two-thirds that of the circum-
scribing right cylinder, and the area of its surface four times
the area of a great circle of the sphere. His request was

—to be described when we come to the seventeenth century. In this connection
see W. H. Bussey, "Fermat's Method of Infinite Descent," *American Mathematical
Monthly*, XXV (1918), 333–337. If Euclid had imagined mathematical induction, as
he might well have done, the fundamental theorem would have been within his
reach. For Zermelo's proof and further references, see G. H. Hardy and E. M.
Wright, *An Introduction to the Theory of Numbers* (1st ed., 1938), pp. 21–22.

honored, and Cicero (106–43 B.C.) in 75 B.C. actually found a commemorative shaft, crumbling and overgrown with brambles. A figure could still be made out, but the accompanying legend was partly obliterated. Archimedes had been dead for 137 years, and the Syracusans had forgotten him, his work and his monument. Cicero concludes his account with a tribute by implication to his Roman home town and, indirectly, to himself: "Thus one of the noblest cities of Greece, which had once been celebrated for learning, would have known nothing of the monument of its greatest genius had it not been disclosed to them by a native of Arpinum." This is charming, as is the rest of the distinguished Roman orator's account of his find. He might have added that this "greatest genius" was butchered by a Roman soldier.

It may seem ungracious (or impertinent) to point out that Archimedes *proved nothing* in the masterpieces mentioned. He was as far from the modern concept of proof as an ancient Egyptian, a pure visual intuitionist. As for the integral calculus, open any up-to-date text and compare it with what the Greeks, including Archimedes, swallowed without noticing a difficulty. Nevertheless they did reason after their own fashion, which was good enough till the nineteenth century. To get an estimate of what Archimedes based his "reasoning" on, see T. L. Heath, *The Works of Archimedes*, Cambridge, 1897.

It would take us too far off our road to describe the work of Archimedes in applied mathematics, but some of it may be mentioned before we pass on to his problem in the theory of numbers. He founded and developed the science of hydrostatics, and his theory of levers made extensive applications to statics. In mechanical invention he was as ingenious as an American. When, during the Second Punic War, the Romans under the consul Marcellus first besieged, then blockaded Syracuse from the sea, the "engines" devised by Archimedes terrified the enemy, if nothing more, and kept him out for three years. The classic yarn about Archimedes setting fire to the Roman fleet by using lenses or mirrors to concentrate the sun's heat is physically impossible and hence nonsense. Yet the tale persists in the popular histories of mathematics. The city fell in 212 B.C. to what Americans since Pearl Harbor have smugly called a "sneak attack" (because it caught our high brass with their hats off) from the rear. Contrary (it is said by apologists) to the express orders of Marcellus, Archimedes, then about seventy-five years old and unarmed, was murdered in the mopping-up operations. The fullest account of all this is in Plutarch's *Lie of Marcellus*. What Marcellus may have wanted with the venerable mathematician, provided he were taken alive, Plutarch does not say. Possibly the baffled but belatedly victorious commander was far ahead of his times, and indeed almost abreast of our own. A man capable of military prowess such as Archimedes had shown was a more valuable prize of war than all the gaudy loot and women in Syracuse.

So much for the official version (Plutarch's) of the death of Archimedes. The unofficial version takes Marcellus down a peg or two. It is due to S. A. Goudsmit of Brookhaven National Laboratory, and was published in the *Proceedings of the American Philosophical Society*, February 1956:

Archimedes was killed by a Roman soldier of the forces that captured Syracuse. General Marcellus apologized profusely for the great mistake. Who knows, however, whether that soldier didn't really belong to an Intelligence Task Force with strict instructions to destroy the enemy's scientific strength, exactly as in the case of our unfortunate destruction of the Japanese cyclotron in 1945.

The American commander-in-chief responsible for this futile barbarity was General ("I shall return") Douglas MacArthur.

Archimedes is credited by tradition with only one contribution to the theory of numbers. This is the famous Cattle Problem, which probably will never be fully disposed of in the sense of producing the complete numerical solution.[6] The circumstances under which the problem came to light in modern times are of interest. The German critic and dramatist, G. E. Lessing (1729–1781), spent most of the last decade of his laborious life dusting and sifting the treasures of the Wolfenbüttel Library in northern Germany, in which Leibniz had slaved out his last unrewarded years. Lessing was chief librarian. He made many discoveries of at least scholarly interest for their time and published them in a series of learned volumes. The first (1773) contained the Cattle Problem, a Greek "epigram" in 44 lines of verse, headed "A problem which Archimedes found among [some] epigrams and sent, in his letter to Eratosthenes of Cyrene, to be solved by those in Alexandria who occupy themselves with such matters." The problem may have been suggested by a passage in the twelfth book of Homer's *Odyssey*: "Next, you will reach the island of Thrinacia [Sicily] where in great numbers graze kine and the sturdy flocks of the Sun."

The complete problem is in two parts. The first is easy today, the second is not, even today. Somewhat shortened, the statement is as follows:

Compute, O Friend, the host of the oxen of the Sun, giving thy mind thereto: if thou hast a share of wisdom, compute the number which once grazed upon the Sicilian isle Thrinacia, and which were divided according to color into four herds, one milk white,

---

[6]See the article by R. C. Archibald, *American Mathematical Monthly*, XXV (1918), 411–414, which contains numerous references to the extensive literature on the problem.

one black, one yellow, and one dappled. The number of bulls
formed the majority of the animals and the relations between
them were as follows—

As we would continue today, let W, w be the respective numbers of white bulls and white cows, and (X, x), (Y, y), (Z, z)
the numbers of bulls and cows in the black, yellow, and
dappled herds respectively. The first part of the problem is to
solve the following seven equations between the eight unknowns:

$$W = \left(\frac{1}{2} + \frac{1}{3}\right) X + Y,$$

$$X = \left(\frac{1}{4} + \frac{1}{5}\right) Z + Y,$$

$$Z = \left(\frac{1}{6} + \frac{1}{7}\right) W + Y,$$

$$w = \left(\frac{1}{3} + \frac{1}{4}\right) (X + x),$$

$$x = \left(\frac{1}{4} + \frac{1}{5}\right) (Z + z),$$

$$z = \left(\frac{1}{6} + \frac{1}{8}\right) (Y + y),$$

$$y = \left(\frac{1}{6} + \frac{1}{7}\right) (W + w),$$

Notice that the fractions are expressed in "unit fractions" as
an Egyptian of the time of Ahmes of the Rhind Papyrus, about
1650 B.C., might have done; thus $\frac{1}{2} + \frac{1}{3}$ instead of $\frac{5}{6}$, which was
not in his arithmetic. These equations are (today) an easy

149

indeterminate system—there are only seven equations for the eight unknowns. The solutions, where n is any number, are

$$W = 10366482n, \qquad w = 7206360n,$$
$$X = 7460514n, \qquad x = 4893246n,$$
$$Y = 4149387n, \qquad y = 5439213n,$$
$$Z = 7358960n, \qquad z = 3515820n.$$

Even for $n = 1$, giving the smallest numbers satisfying the equations, the herd must have stood several deep on the island of Sicily. But this is only a paltry beginning. In passing, the calculations for this much of the problem, simple to us, might have taxed a skilled computer having at his command only the Greek alphabetical system of writing numbers or even the Babylonian sexagesimal. Archimedes may have suspected as much, for he says, "If thou canst give, O Friend, the number of bulls and cows in each herd thou art not unknowing nor unskilled in numbers, but still not yet to be counted among the wise." He then stiffens the problem very considerably by imposing two further conditions which the numbers of bulls must satisfy.

Consider, however, the following additional relations between the [number of] bulls of the Sun:

$$W + X = \text{a square number}$$
$$Z + Y = \text{a triangular number.}$$

When thou hast computed the totals of the herds, O Friend, go forth as a conqueror, and rest assured that thou art proved most skilled in the science of numbers.

That was rather rubbing it in. With the values of W, X, Z, Y, as above satisfying the first part, the problem now demands the solution of

$$W + X = u^2, \qquad Z + Y = \tfrac{1}{2}v(v+1),$$

where u, v are unknown numbers, and this in turn, as easily shown, requires numbers T, U to satisfy

$$T^2 - 4729494U^2 = 1.$$

The last equation is of the general type

$$T^2 - DU^2 = 1,$$

where D is a number having no square divisor exceeding 1. By a historical mistake, it is named after an interesting character but mediocre mathematician, John Pell (1610–1685), and is called a *Pellian equation.*[7] It is too late now to right the mistake; the equation should have been named after Fermat. We shall meet the general type again when we come to Fermat and his contemporaries. The method for solving it completely is now well known and is the powerful algorithm of continued fractions in the current theory of numbers. The older English school algebra used to include the algorithm. We shall come back to this in connection with Fermat. For the present, it is by no means obvious that an equation of this type *for any given D of the prescribed form necessarily has a solution T, U, and indeed an infinity of solutions.* The *existence* of a solution was proved only in 1776, when Lagrange (1736–1813) succeeded after many attempts which, he said, had cost him more thought than any of his other great successes and perhaps more than they may have been worth.

---

[7]Its appearance in India was remarked in Chapter 4.

The smallest solution T, U, for Archimedes' $D = 4729494$ was computed in 1880 by A. Amthor. It is $T = 109,931, 986,732,829,734,979,866,232,821,433,543,901,088,049$, $U = 50, 549,485,234,315,033,074,477,819,735,540,408,986,340$.

I gladly leave to the reader the pleasure of verifying that these T, U actually satisfy the equation. If—which seems most unlikely—Archimedes knew that his final equation must have a solution, he was about twenty centuries ahead of his time. In fact he would have had to know as much as Lagrange unless, of course, he succeeded in calculating the required numbers T, U —an extremely laborious process—and proceeded thence in an attempt to compute the total number of cattle in the herd. If he succeeded in the last he surpassed any calculating machine yet invented or ever likely to be. For Amthor proved that W is a number of 206545 digits, and the total number in the herd also requires this many. To give some idea of the magnitude of these numbers, Amthor said, "It is easy to show that a sphere having the diameter of the Milky Way across which light [speeding at 186,000 miles a second] takes ten thousand years to travel, could contain only a part of this great number of animals even if the size of each is that of the smallest bacterium." Amthor gives another illustration. To print all eight numbers of the solution, with 2500 digits to the page, would require a volume of over 660 pages. After all this it seems improbable, to say the least, that Gauss solved the Cattle Problem completely, as one of his uncritical admirers asserted that he had. There must have been a misunderstanding somewhere.

A cynic has remarked that as long as there is an unsolved problem, some fool will try to solve it, especially if the problem is unsolvable. Perhaps A. H. Bell (no relation to the present writer) and his two collaborators, constituting the Hillsboro Mathematical Club of Hillsboro, Illinois, in the 1890s, did not fully qualify for the cynic's remark, but they came close. After nearly four years of unpaid hard labor the members of the club computed 30 or 31 of the first digits and 12 of the last for each of the eight unknown bull and cow numbers and for the total number in the herd. The results disagree with the more modest computations by Amthor. Does anybody wish to check the club's computations? Or Amthor's? The field is open and wider than the plains of Thrinacia.

There is the inevitable question, was it really Archimedes who proposed the problem? If it was not, who else had brains enough to imagine such a horror? The historical experts agree that Archimedes probably was responsible for the first part. Some, but not all, credit him also with inventing the second part. Again if it were not Archimedes, what unknown genius constructed an elaborate solvable problem involving indeterminate equations of the *second degree?* Equations of this kind written down at random are far more likely than not to have no solution. The final verdict, according to some scholars, is that it was Archimedes after all who successfully set the hard second part. The evidence is not mathematical but human:

The unmistakable vein of satire in the opening words of the epigram, and in the transition from the first to the second part, and in the last lines, was a shaft directed toward Apollonius.[8]

This is the only time that Apollonius, "the Great Geometer" as he was justly called by the Greeks, the Moslems, and their successors, figures as a contributor to the theory of numbers. A geometer who could inspire anything as fiendish as the Cattle

[8]T. L. Heath, *Diophantus of Alexandria* (Cambridge: Cambridge University Press, 1910), p. 122.

Problem must have been an arithmetician, even if only an involuntary one, of no mean stature. If Archimedes ever sent his respected friend Eratosthenes anything for himself nearly so provocative as the Thrinacian bulls and cows, it has not survived.

The occasion for Archimedes' "shaft" was the problem of approximating to π (pi), the ratio of the circumference of any circle to its diameter. To seven places, $\pi = 3.1415927$, which is considerably farther than any of the Greeks got. (In 1949 a modern calculating machine in about seventy hours [the time was later reduced to thirteen *minutes*] performed the feat of computing π to 2035 decimals.[9] Archimedes had shown that π is less than $3\frac{1}{7}$ and greater than $3\frac{10}{71}$. Apollonius claimed a closer approximation, and is said to have boasted about his superiority over Archimedes. He also promoted a system for writing and manipulating large numbers that he claimed was better than the similar venture of Archimedes in his "sand reckoner"—which need not be described as it has nothing to do with the theory of numbers; it was a scheme, ingenious for its time, of constructing large numbers. Archimedes drove Apollonius to cover by hurling the Cattle Problem at him from ambush. The huge numbers required to solve the problem were beyond human computation by any means Apollonius could possibly have known. They might even cause an electronic calculator to blow a battery of tubes.

I have mentioned the indirect slur on Apollonius' character and personality by Pappus. The alleged conceit and the reputed ungenerous treatment of his predecessors and contemporaries are not evident in Apollonius' masterpiece, his treatise on conic sections. As was his plain right, he pointed out

---

[9] Recently some enthusiastic idiots computed π to ten thousand decimals.

wherein his work differed from that of his predecessors, and stated what he claimed as original with himself. Archimedes did no less. Apollonius' geometrical method was in the Greek purely synthetic tradition, like Euclid's, and was not superseded till the seventeenth century, when Descartes' analytic method, published in his *Geometry* of 1637, quickly ousted the other. As a synthetic geometer, Apollonius was without a rival till the nineteenth century, when the Swiss, Jakob Steiner (1796–1863)—"the greatest [pure] geometer since Euclid"—resumed and extended the old way in a fashion that would have astonished and might have pleased Apollonius.[10]

Almost nothing is known of Apollonius' life. About twenty-five years younger than Archimedes, he was born in 260 B.C. at Perga in Pamphilia, a town still remembered in connection with St. Paul and Barnabas on their first mission to Asia Minor. As a young man he went to Alexandria, where like Archimedes he studied under the successors of Euclid. Except for short journeys, he probably lived out his productive years at Alexandria.

With the unequaled triumvirate of Euclid, Archimedes and Apollonius, Greek mathematics reached its greatest height. After the incomparable three, with the one exception of Diophantus, there was a rapid and steep decline. Diophantus is not due yet on the historical scene; we must first meet Eratosthenes and Nicomachus.

Eratosthenes (275–194 B.C.), to whom Archimedes sent the Cattle Problem, received his education in Athens and moved to Alexandria in his thirties. His nickname, Pentathlos, mean-

---

[10]Steiner could not read till he was fourteen, so if your boy is slow in learning to read, he may even so become a great geometer.

ing an athlete who competed at the Olympic Games in five of the sports, jumping, foot racing, wrestling, throwing the discus, and hurling the javelin, is a wry compliment to his abilities in many things and a jibe at his failure to be first in any of them. However, some of his seconds would have been first in any epoch of antiquity that did not include such champions as Archimedes and Apollonius. Archimedes had the highest regard for him, as both a man and a scientist, and confided to him his revealing work *On Mechanical Theorems, Method*. In this Archimedes explained how he had used his mechanics as a heuristic guide to results in pure mathematics. (*The Method*, long lost, was found in Constantinople in 1906.) Among Eratosthenes' high seconds was his determination of the diameter of the earth. His estimate was only about fifty miles off the correct polar diameter, a most remarkable achievement for its time. He was less successful in his attempts to measure the sun and the moon—the sun came out only twenty-seven times larger than the earth. He also wrote a commentary on Plato's *Timaeus*, a work on mathematical geography, and the beginning of systematic chronology. His unique contribution to the theory of numbers, described presently, is the only one of all his many works that has retained its interest. With his insatiable curiosity and vast erudition, Eratosthenes was the ideal candidate to succeed the poet and cataloguer Callimachus as director of the Alexandrian Library. When he was about forty he became the third chief librarian in order of succession, and found himself buried in more books than even he could hope ever to read his way out of. When his eyesight failed and he could no longer read or do mathematics, he committed suicide.

Eratosthenes is remembered in the theory of numbers for his sieve, a simple device for sifting out the primes from the sequence of integers. The following, without any of the current modifications, is the form in which Eratosthenes gave it. Beginning with 2, which is a prime, write down all the numbers up to some prescribed limit, say 100, or imagine the sequence continuing indefinitely,

$$2, 3, 4, 5, 6, 7, \ldots, 98, 99, 100, \ldots$$

Start with 2 and mark every second number after 2. This marks the multiples of 2. The smallest unmarked number after 2 is 3; mark every third number from 3 on. (Some numbers will now be marked twice, for example 6.) The smallest unmarked number after 3 is 5; mark every fifth number from 5 on. Continue thus with the smallest unmarked number at each stage. These smallest unmarked numbers are the primes. Some critics have called the sieve a trivial and feeble device, without themselves suggesting anything better. One even suggested something much worse. Actually it has been used for making stencils and other mechanical aids in the construction of factor tables.

As noted at the beginning of this chapter, the date of Diophantus has now been temporarily moved back to the first century A.D. If this is correct (it may not be; see Chapter 7 here) it raises an awkward question about Nicomachus in the same century. Was he acquainted with the work of Diophantus? Whatever the answer, it is of no great consequence here. Nicomachus was such a muddle-headed mystic that anything he might have lifted from Diophantus would be smothered in Neo-Pythagorean metaphysics by the time he had finished with it. Nevertheless, Nicomachus must be considered on the road from Babylon to Fermat's France. Compared to Diophantus, Nicomachus is a pigmy. But he is important for two reasons: he is a conspicuous example of the precipitous decline after Euclid, Archimedes, and Apollonius; he transmitted some amusing things about polygonal numbers, and so was partly responsible for the remembrance of these numbers in Fermat's seventeenth century. These items may not add up to much, and perhaps mathematical readers will wonder why Nicomachus is mentioned at all. They will see if they persevere to the report of his canonization.

Nicomachus (first century A.D.) was a native of Gerasa, a small town about fifty miles northeast of Jerusalem. A Neo-Pythagorean, he wrote on music in the Pythagorean tradition, and may have put together a compendium on the rudiments of elementary geometry. But it is by his "masterpiece," such as it is, the *Introduction to Arithmetic*,[11] that he survives, if only as an example of how *not* to do mathematics. The *Introduction* was aimed at philosophers, which may account for its watered-down arithmetic and its intolerably verbose rhetoric. As normal philosophers are usually allergic to real mathematics, the book had an immediate and protracted success. Had it been read only by those to whom it was primarily addressed not much harm would have been done. But for over a thousand years this travesty of mathematics was the accepted authority on the properties of numbers. The Christian martyr Boethius (died 524) perpetuated some of it in his Latin text on arithmetic, thereby debasing mathematical instruction in Europe for centuries. The high esteem in which Nicomachus was held in the Middle Ages and earlier—also in some scholarly quarters today—appears in the ludicrous misunderstanding of a sour compliment the Greek satirist Lucian (in the second century) paid a certain calculator: "You reckon like

---

[11]*Introduction to Arithmetic*, translated into English by Martin Luther D'Ooge, etc. (Ann Arbor: University of Michigan Press, 1938). Anyone who may imagine that the estimate here of Nicomachus' masterpiece is too harsh or unjust, is invited to read it in this excellent translation.

pion proposer and solver of hard arithmetical problems was spared the humiliation of witnessing this puerile, and frequently quoted, puzzle sponsored in his name.

Diophantus lives entirely in his work. His name survives in the technical term *Diophantine analysis*, which is the art of solving indeterminate equations in *whole numbers* or in *rational numbers* (common fractions). To explain the very simple technicalities: an equation is called *indeterminate* if it admits an infinity of solutions, and likewise for a simultaneous system of two or more equations. For example, x, y being the unknowns, $2x + y = 1$ is indeterminate. A *rational number* is the "ratio" of two integers, whence the name. Thus a rational number is of the form m/n, where m, n are integers. Without loss of generality m, n in m/n may be considered coprime, as any common factor could be canceled, as in $21/98 = 3/14$. A "number" which is not rational is called irrational. For example $\sqrt{p}$, the square root of any prime p, is irrational because there are no integers m, n such that $p^2 = m/n$. ($m = p^2$, $n = 1$ gives $p^2 = p^2$.)

Diophantus and his successors for centuries were unacquainted with negative numbers, although it has been argued from such evidence as

$$(6 - 2) \times (5 - 3) = 4 \times 2 = 8,$$

that he knew the rules "Minus times minus is plus, minus times plus is minus, plus times minus is minus, plus times plus is plus" as in

$$(6 - 2) \times (5 - 3) = (6 \times 5) - (6 \times 3) - (2 \times 5) + (2 \times 3)$$
$$= 30 - 18 - 10 + 6 = 8.$$

163

His lack of negative numbers often made his problems harder to solve than they would be today, when positives and negatives are usually accorded equal status as numbers. On the other hand his *admission of rational numbers* (*common fractions*) *as solutions* made many of his problems much easier than if he had demanded *integer* solutions, as is frequently the custom today, and indeed has been since the seventeenth century. But even with all the latitude he allowed himself, most of the indeterminate problems he solved with unsurpassed ingenuity are hard enough.

We shall need three more technical terms, *degree*, *homogeneous*, *inhomogeneous*, especially as applied to equations in the present context. The *degrees* of the successive powers $x, x^2, x^3, x^4, \ldots, x^n$ are respectively, $1, 2, 3, 4, \ldots, n$, and it is convenient sometimes to speak of these as being of the first, second, third, fourth, $\ldots$, nth degrees. The degree of a product is defined to be the sum of the degrees of its factors. Thus $x^2 \times x^5$ is of degree $2 + 5$ or 7 in x; $x^3 y^2$ is of degree $3 + 2$ or 5 in x and y jointly. Diophantus knew and used these notions, which in his day were far from commonplace. An orthodox geometer in the Euclidean tradition would have been horrified by such a multiplication as $x^2 \times x^3$, for to him x^2 would represent a square (an area), and x^3 a cube (a volume), and what inconceivable monstrosity would spring from the multiplication (marriage, as a Pythagorean would have said) of an area by a volume? Such profoundly meaningless difficulties bothered Diophantus no more than they do us, and just as we should, he *added the exponents* $2, 3$ in $x^2 \times x^3$ to get 5, giving him $x^2 \times x^3 = x^5$. This made sense, as he was dealing with *numbers* and had abandoned geometrical imagery for arithmetical language. Geometrically, such an expression as $x + yz$

where x, y, z represent line segments, would signify the meaningless operation of adding a line to a rectangle. To the classical Greeks, geometrically, this was a senseless abomination. For Diophantus it was merely the addition of two numbers.

Equations are classified today according to their degrees. It was not clear in the early development of algebra that this classification, as algebraists later found out, is the productive thing to do. The degree of an equation is defined as the greatest degree of any term in it, the "terms" being the various powers and products of the unknowns occurring. For example, if x, y are unknowns to be found from the equation $2x^3 + 3xy^2 = 5$, the degree is 3, because each of the terms x^3, xy^2 is of degree 3: the equations $x^5 + 5x^3 + 2x + 1 = 0$, and $x^2y^3 + 7y = 2$ are of degree 5.

An equation is called *homogeneous* if all the terms in it are of *the same degree* in the unknowns; if terms of *different degrees* in the unknowns occur, the equation is *inhomogeneous*. Thus the first of the following is homogeneous of degree 3, the second is inhomogeneous of degree 4 in the unknowns x, y: $4x^3 + 2x^2y + y^3 = 0$; $x^4 + 3x^3y + y^3 + 2 = 0$.

All this detail is necessary because in Diophantine analysis there is a fundamental difference between homogeneous and inhomogeneous equations. The problem of finding the *rational* solutions of a homogeneous equation is the same as that of finding the integer solutions; for inhomogeneous equations the two problems are different, and the solution in integers is harder than that in rational numbers (see concluding chapter). Diophantus considers both types of equations in his numerous problems, and seeks to solve them in *rational numbers*. By not imposing the restriction that the solution be in *integers*, he softened things up very considerably for his inhomogeneous

165

equations. Still, many of them are sufficiently difficult even today, and some are inaccessible if, as is customary now, *all* solutions are demanded.

Before sampling Diophantus' problems, we may consider his standard technique for solving many of them. First, he is content with a *single numerical solution*. This might seem to put him on the same level as Nicomachus. But there is a difference. Had Diophantus followed Nicomachus, he would merely have exhibited the required numbers without even hinting how he got them. But he has a method all his own. He assumes a particular number as a solution, and tests it. If it satisfies his equation, he is through. If it does not, his work may show him why, and he then modifies some detail sufficiently to enable him to get around the obstacle. For example, if in his first attempt to find a rational solution for some equation of the second degree he runs into a square root, say $\sqrt{a^2 + bc}$ where a, b, c are rational numbers, he chooses a, b, c so that the square root is a rational number. Thus he might choose $b = r + a$, $c = r - a$, where r is a rational number greater than a (since he operates only with positive numbers), which would give r for the square root. But if he cannot get rid of an irrational number appearing in the attempted solution of a particular equation, he goes back, modifies the tentative solution in a manner suggested by the form of the obstacle, and makes a fresh start. These steps may not be evident in the final solution, and sometimes we can only try to puzzle out what made him propose some special problem instead of some other closely like it. Doubtless he had sufficient reasons. And although he drops a problem when he has found a single rational solution, his technique is usually capable of furnishing either further solutions or new solvable equations from a given one.

His superb skill in choosing numbers that enable him to produce solutions to his equations marks him as the master tactician of numerical analysis.

Diophantus composed three works on the theory of numbers: *Arithmetica* in thirteen books, of which only six survive; *Porisms*, all lost; a treatise on *Polygonal Numbers*, lost or incomplete except for an unimpressive fragment. The last contains nothing of interest;[15] the word *porism* is practically obsolete in mathematics, as is also the type of problem to which it referred in Greek mathematics. I transcribe the dictionary definition: "A proposition affirming the possibility of finding such conditions as will render a certain problem indeterminate, or capable of innumerable solutions." Such propositions would obviously have been of importance to Diophantus, and he refers in the *Arithmetica* to his own missing porisms for justification of certain assertions.

The *Arithmetica* is the treasure chest of Diophantine analysis in the ancient tradition of its originator—that of finding *particular rational solutions* of indeterminate equations. So far as is known, this work was Diophantus' masterpiece. It almost shared its author's eclipse, and but for a series of improbable,

---

[15]The fragment concludes with an abandoned attempt to find the number of ways in which a given integer is a polygonal number (see Chapter 3 here). Heath,[8] p. 259, reproduces an illusory scheme by G. Wertheim (1897) for finding the required number of ways. See also Dickson's *History*, II (1920), 3 for Wertheim's scheme. It comes nowhere near to solving the problem, because the essential step requires *all* resolutions of *any* integer into a pair of factors or, what is equivalent, finding all the divisors of any given integer. What Wertheim offers is thus a purely tentative process, that might or might not succeed in a reasonable time—say a hundred years—in a particular instance, even with calculating machines. Not even the most powerful machine yet invented could crack a really large integer represented in the decimal scale by a string of random digits supplied by another machine. So Diophantus' problem is still open, and likely to remain so for a long time; it may even survive Fermat's Last Theorem.

but true, accidents might be as completely lost as he himself. In the following chapter we shall see how the surviving fragment found its way into the hands that could make the most use of it, Fermat's. For the moment we must note briefly the general character of the surviving six books. I shall describe only some of what concerns indeterminate equations, restating Diophantus' verbal problems in current symbolism. The letters a, b, c, d, e, f stand for *given* positive rational numbers, and x, y, z for *unknowns* to be found as positive rational numbers.

Diophantus, being content with *rational* solutions ignores *indeterminate* equations of the *first degree*, such as

$$ax + by = c,$$

since these, in his philosophy, are trivial. In the specimen, $x = (c - by)/a$ is the solution where y may be any rational number which makes by less than c.

Most of his equations are of the *second* degree, and several are equivalent to special cases of

$$y^2 = a^2 x^2 + bx + c.$$

Note a^2, which makes the problem easy—in modern symbolism. The appropriate device is evident: get rid of the term in x^2 by replacing y by $ax + z$, where z is a new unknown. The result is the new equation

$$2axz + z^2 = bx + c,$$

which is of the *first* degree in x, and so is immediately solvable for x in terms of z: $x = (c - z^2)/(2az - b)$; whence since $y = ax + z$, we find y. It remains only to choose for z any rational number which will make these x, y positive. Although for any particular a, b, c Diophantus was content with a single rational solution, his procedure is essentially that above and, like it, capable of furnishing any number of solutions x, y. All he lacked was our algebraic notation. He had the beginnings of a notation himself, as we shall see presently.

Another frequent type is

$$y^2 = ax^2 + bx + c^2.$$

Note c^2. This type is disposed of by the substitution $y = zx + c$.

He also solves certain simultaneous equations of degree 2. If there are two equations he calls the system "double," a locution that lasted through the time of Fermat. Two types occur:

$$y^2 = a^2x^2 + bx + c, \qquad z^2 = d^2x^2 + ex + f;$$

and

$$y^2 = ax^2 + bx + c, \qquad z^2 = dx^2 + ex + f.$$

After the devices explained for single equations it is clear how to proceed. Essentially the same device applies to his equation of degree 3,

$$y^3 = a^3x^3 + x^2 - a^3x - 1.$$

Both a^3x^3 and -1 can be got rid of on replacing y by $ax - 1$ and reducing the resulting equation. This gives an equation of degree 1 for x.

A more interesting type of problem takes us back through Plato and Pythagoras to Babylon: to find certain right triangles whose sides, for example, satisfy prescribed conditions. To solve these, he proceeds from a solution of $x^2 + y^2 = z^2$.

A few of the harder problems (restated in current notation) must suffice to give some idea of his ingenuity and versatility.

169

As always the solutions are to be rational numbers. All letters denote unknowns unless otherwise stated.

$$xy + x = u^2, \qquad xy + y = v^2, \qquad u + v = a,$$

in which a is a given number.

$$x^2 + x + y + z = u^2, \; y^2 + x + y + z = v^2, \; z^2 + x + y + z = w^2.$$

$$xy + z = u^2, \; yz + x = v^2, \; zx + y = w^2.$$

$$xy + z^2 = u^2, \; yz + x^2 = v^2, \; zx + y^2 = w^2.$$

$$x^2 + xy + y^2 = u^2.$$

$$x^2 y^2 z^2 + x^2 = u^2, \; x^2 y^2 z^2 + y^2 = v^2, \; x^2 y^2 z^2 + z^2 = w^2.$$

$$xy + x = u^3, \; xy + y = v^3.$$

If we had to find *all* the solutions of certain of these equations we should be baffled. Some of them exercised Euler and others in the eighteenth century. In the tradition of Diophantus most of those who attacked equations of this difficulty settled for particular solutions.

I note specially Problem 29 in Book 5,

$$x^4 + y^4 + z^4 = w^2,$$

because it suggested to Fermat the equation

$$x^4 + y^4 = z^2,$$

which he proved impossible. From this followed the impossibility of $x^4 + y^4 = z^4$. Diophantus' problem is of current interest for another reason. It is easily shown that his equation has an infinity of integer solutions[16] x, y, z, w. Going up a step, we might try to solve

$$x^4 + y^4 + z^4 = w^4.$$

Whether this equation is solvable or not is an unsettled question. Euler said he thought the equation is impossible. If Fermat tried it he does not say so. Should anyone wish to explore for a possible solution he may start with integers w greater than 10,000, for up to that, as shown by Morgan Ward, there is no solution.

In addition to indeterminate equations, Diophantus has a few general theorems on numbers which at least are remarkable for their epoch. For example, he states that any given square number is expressible as a sum of two squares in any number of ways. The problem is to solve $x^2 + y^2 = c^2$, in *rational* numbers x, y, where c is any given rational number. One method is immediate. Replace y by $xz - c$, and find $x = 2cz/(z^2 + 1)$, whence $y = c(z^2 - 1)/(z^2 + 1)$, so that z may be any rational number greater than 1. Again, he states that any sum of two cubes is also a difference of two cubes. He knew that no number of the form $8n + 7$ is a sum of three

---

[16] For example the following identity gives an infinity of integer solutions for integer values of a other than 1 or -1,

$$(2a^3 - 2a)^4 + (2a^3 + 2a)^4 + (a^4 - 1)^4 = (a^8 + 14a^4 + 1)^2.$$

For $a = 2$, $12^4 + 20^4 + 15^4 = 481^2$,
for $a = 3$, $48^4 + 60^4 + 80^4 = 7696^2$,

$$(2^4 \times 3)^4 + (2^2 \times 15)^4 + (2^4 \times 5)^4 = (16 \times 481)^2$$
$$2^{16} \cdot 3^4 + 2^8 \cdot 15^4 + 2^{16} \cdot 5^4 = 2^8 \cdot (481)^2$$
$$2^8 \cdot 3^4 + 15^4 + 2^8 \cdot 5^4 = 481^2.$$

squares. He gave an equivalent of the algebraic identity,

$$\left(a^2 + b^2 \right)\left(c^2 + d^2 \right) = \left(ac + bd \right)^2 + \left(ad - bc \right)^2,$$

and the like with the signs $-$, $+$ instead of $+$, $-$ on the right. From this it follows that a product of two integers, each of which is a sum of two square integers, is a sum of two square integers. (To satisfy Diophantus' constant limitation that only positive numbers are recognized ad and bc must be unequal in the first statement, and ac and bd in the second.) This identity reappears in the thirteenth century with Leonardo of Pisano (Fibonacci), and was the starting point in the nineteenth and twentieth centuries for much work concerning products of sums of squares. The reader who remembers some trigonometry will observe the analogies with the addition theorems for the sine and cosine. Finally, Diophantus may have noticed that some integers are squares or sums of not more than four squares. The general theorem that any integer is of this form was first proved in 1772 by Lagrange. Several proofs have been given; all are either difficult or artificial. The theorem is of interest to us here because it started Fermat toward one of his finest results—that on polygonal numbers, which I have described in Chapter 3.

One secret of Diophantus' success is his skill in using the rudimentary algebra he invented. Although little more than a scheme of abbreviations for key words, this proto-algebra was a long advance beyond the "geometrical algebra" of Euclid as in Book 2 of his *Elements*. Diophantus symbolized powers of the unknown up to the sixth; he also symbolized the reciprocals of these powers. There was a symbol for the unknown which, appropriately enough, was called the "number" (*arithmos*). A troublesome lack was that of different symbols for more than one unknown, but with his usual ingenuity he got around this.

His head was phenomenally clear. He had a special sign for minus, but none for plus; addition of terms was indicated by juxtaposition.

It remains to say something about his *determinate* equations such as those in a first course in algebra today. The most interesting historically are those in which the sum, or the difference, and the products of two numbers are given, and it is required to determine the numbers. These go back to Babylon. As an example, he solves the system $x + y = 2a$, $xy = b$, for given numbers a, b, which he reduces to a quadratic equation solvable within his restrictions. If such an equation has two positive roots we might expect him to produce both, but he is content with one. Naturally when an equation presents him with what in his philosophy is an impossible square root, as does his $4x^2 + 20 = 4$, which we would simplify by dividing out 4 to get $x^2 + 5 = 1$, or $x^2 = -4$, he dismisses the equation as the absurdity it is.

In addition to the unsettled date of Diophantus there are two questions the historians would like to have answered. Was Diophantus a Greek? If so, he was a sport. He does not think like any of his Greek predecessors, nor is the kind of problem he attacks one that interested them. Again, was his *Arithmetica* all his own, or was it, in the manner of Euclid's *Elements*, at least partly a compilation from earlier works? There has been much inconclusive argument on both questions. Possibly he was just a mathematical genius with new ideas. There have been such. Actually, some of his problems occur in Babylonian mathematics. The route by which they reached him has still to be traced.

From the matter-of-fact Diophantus to "the divine Iamblichus," who died about A.D. 330, is quite a long step down. Iamblichus is remembered today chiefly because his

173

account of the Pythagorean sect and its founder is a store of information, and misinformation according to some, concerning the Pythagoreans. Iamblichus was born at Chalcis in Syria, and apparently was a man of some wealth. He has the questionable honor of having founded the second, or Syrian, school of the Neoplatonic philosophy, in which all the numerology of the Pythagoreans and their successors, including Plato, was combined with Oriental mysticisms, universal theogonies, and miscellaneous superstitions into one stupendous confusion. Number-nonsense united the conflicting theologies of a swarm of deities in an inconsistent parody of consistency beyond all logic. This was the supreme effort of pagan numerology. Iamblichus had no worthy successor till the early Christian theologians and their successors in the Middle Ages went on where he left off.

Iamblichus is credited with a work in nine books, *Collection of the Pythagorean doctrines*, of which about half survives in *The Life of Pythagoras*; *Exhortation to philosophy*; *Concerning the mathematical science in general*; *Introduction to arithmetic* (after Nicomachus); *Arithmetical theology*. It would be interesting to know how much of the first is to be believed, composed as it was about eight centuries after the presumed death of Pythagoras. Even if it may be mostly lies, it is at least not as dull as some supposedly truthful histories.

In the matter of perfect numbers, Iamblichus went infinitely far beyond the "fair and excellent things" of Nicomachus who, as we saw, stated (without bothering to prove his assertion) that there is one and only one perfect number in the respective intervals between 1 and 10, 10 and 100, 100 and 1000, 1000 and 10,000. Iamblichus says this rule holds "indefinitely"—for the intervals between 10,000 and 100,000, 100,000 and 1,000,000, ... and so on, forever. He also says that all the

7

Dating—Collapse—Recovery

1. Dating: Pythagoras, Diophantus

"The majesty of a great river does not efface the charms of its humble source."[1]

The "great river" is the theory of numbers as it exists today; the "humble source" is the work of Diophantus. One mild protest: the *Arithmetica* of Diophantus could hardly be called humble; it was a major effort of high mathematical genius. But let that pass, the meaning is clear. The course of the river is extremely complicated, especially in its earlier meanderings. To follow it at all in a reasonably limited space, I can note only a few of its more prominent bends and turns from Diophantus to Fermat.

A strict chronological order from Hypatia's Alexandria to Fermat's France is impossible, as we must make short and unimportant detours of both time and place to India, Arabia,

---

[1] Translation from Paul Ver Eecke, *Diophante d'Alexandrie* (Bruges, 1926). N.B.: To avoid needless repetition, references to Heath in this chapter are to his *Diophantus of Alexandria*, cited in Chapter 6; the reference to Neugebauer is to *Archeion*, cited in Chapter 3.

179

Persia, and Byzantium. The river is not a single stream historically, but a confluence of several rivers spreading over many diverse civilizations.

Taking what at first sight may seem an unlikely starting point in time, I shall first return briefly to Pythagoras in the sixth century B.C. (see Chapter 3), drop him there, and pick him up again in the fifth century A.D. in Byzantium, when we will say farewell to him for good. So far as concerns Pythagoras himself (though not the entire Pythagorean cult named after him), the account given in Chapter 3 followed the traditional legends, persistent for centuries. For these there is no documentary evidence. So, historically, Pythagoras is nonsuited and thrown out of court. Neugebauer says: "I do not doubt that any connection with the name of Pythagoras is purely legendary and of no historical value." That is plain enough. But some allusion to historicity and scholarly dating must be made even in an account like this. I do so now, for Pythagoras, Diophantus, and Hypatia's reputed recension of the latter's *Arithmetica*, before proceeding to the mathematical frivolities of the dissolute and carefree Byzantines. Those not interested in the following short hypothetical chronology may pass at once to the next main topic, *Collapse*.

I shall retail first the supposed historical origin of the most significant of all the legends concerning Pythagoras, that crediting him with imposing the axiomatic method on the study of elementary geometry. The documentary evidence, such as it is, now judged "incompetent, immaterial and irrelevant, and not proper cross-examination," on which Pythagoras' claim is based, is quoted (and thus is hearsay) by the Neoplatonic philosopher Proclus (A.D. 412–485) of Byzantium, roughly 900 years—quite a long time—after Pythagoras had died (if such a man ever lived). Proclus wrote a commentary on Euclid's *Elements*, of which only that on Book 1 survives. Like many other Neoplatonists, Proclus occasionally was inclined to be fuzzily mystical. He partly redeemed himself, however, by including in his commentary some alleged history of early Greek mathematics based, supposedly, on the *Summary* by Eudemus of Rhodes (fourth century B.C.). So about 700 years separated Eudemus and Proclus. Eudemus was a competent mathematician—Apollonius respected him. He had been a pupil of Aristotle's (384–322 B.C.), who is said to have suggested that he write a history of mathematics. One excerpt, following Proclus, will suffice:

Pythagoras changed the study of geometry into the form of a liberal education; for he examined the principles to the bottom, and investigated its theorems in an immaterial [abstract] and intellectual [logical] manner.[2]

Several Greek works of the fourth century B.C. have survived; the Eudemean *Summary* has not. All we know of it is what Proclus wrote, say seven centuries after it supposedly was written. It may turn up yet—recall the *Method* of Archimedes mislaid for over 2100 years. In the meantime all we have is hearsay.

---

[2] It used to be said that Thales did what Eudemus attributes to Pythagoras. Thales is credited by tradition with the statements of certain simple geometrical propositions. There is no claim that he proved, or attempted to prove, any of them. The Babylonians had anticipated Thales in several of his stated propositions. As we saw, Babylonian geometry was purely empirical. Eudemus would have been on safer ground if he had claimed the introduction of proof for the Pythagoreans, and not for the legendary Pythagoras himself, who may never have existed.

Compared to Diophantus and Hypatia's reputed recension of his *Arithmetica*, the shadowy and elusive Pythagoras is as fixed and constant as the North Star. Hypatia has one undisputed date, attested by Christians and pagans alike, that of her murder, A.D. 415. It will be interesting to see the kind of evidence, accepted by some famous and respected historians of mathematics as reliable, for the date of Diophantus, and also for the material existence at some unspecified time of Hypatia's recension. Both obviously are important historically for the theory of numbers.

First, Diophantus. *When* did he live? Or, more precisely, *how many years ago did he live*? Such questions do not necessarily lead professional historians to the same answer, as will appear shortly in connection with Christ. After discarding one somewhat wildly guessed date, Heath says "the positive evidence on the subject [D.'s date] can be given very shortly." The gelatinous substance of the offered "evidence" is as follows. Diophantus quotes from Hypsicles [between 200 and 100 B.C.], which fixes the date as later than 150 B.C. Because Diophantus is quoted by Theon of Alexandria, he must have been earlier than A.D. 350. "A letter from Psellus *in the eleventh century* [my emphasis] contains an allusion to Diophantus... ." More of the like divination leads Heath to the conclusion that Diophantus flourished in A.D. 250. "This agrees with the fact that he is not quoted by Nicomachus [about A.D. 100], Theon of Smyrna [about A.D. 130], and Iamblichus [end of the third century]." As for Nicomachus, it is not impossible that he had at least heard of Diophantus' *Arithmetica* but, being the mathematical incompetent and enthusiastic mystic that he was, wisely ignored it to save whatever face he may have had. If all that Heath says is factual and relevant, why has the date of Diophantus been shifted back to the first century A.D.? Or has it? How reliable are the dates of Hypsicles, Nicomachus, Theon, and Iamblichus, accepted by Heath as firmly established? A skeptical young chemist, after analyzing Heath's elaborate and involved argument, said "If you reasoned like that in chemistry you would blow yourself to hell."

An obvious source of the difficulties in historical dating is the lack of any fixed and trustworthy point of reference. There may be one, but if so it has not yet been revealed, at least to mathematicians. In the foregoing, B.C. and A.D. refer of course to the year of Christ's birth, but how many years ago was Christ born? Embittered theologians and "higher critics" have been brawling over this for decades, coming up (sometimes) with answers from thirty to seventy or even a hundred years apart. Some even deny that such a person as Christ ever lived. What is needed is not a date B.C. or A.D., meaningless here, but a reasonably precise number of years reckoned back from the present. For some doubtful dates in history, astronomical records have helped. More recently, carbon 14 and other radioactive elements have been used in dating material remains—mummies, for instance. There was an amusing demonstration of radioactive dating when four mummies and their cases were exposed to the Geiger counters. Three of the specimens accorded reasonably well with the archaeological evidence; the fourth did not. Archaeologists then detected it as a brilliant fake, fabricated (it was said) for sale to an American museum. But nuclear physics could not help us with the date of Christ unless, of course, some certified relic, such as a splinter of the true cross, could be used. The whole difficulty resembles what topologists call the location of fixed points—"the fixed point theorems": in an infinite swirl of possible configurations there are sometimes one or more points that remain fixed under a set of continuous transformations. There is the like in

183

dynamics. A single fixed point in the remoter past of mathematics is needed, but none is in sight, nor likely to be so long as interest in history lasts.

Another source of doubt and confusion can appear in the printed date of a printed book: the author, or the publisher, or both, may have lied. All these have been known to happen, many times. Some have led to lucrative swindles, particularly in the humanities. With current scientific techniques of detection, scholarly frauds are less common than they were as late as the 1890s. In mathematics they have always been rather rare, but correspondingly more interesting. When a book, such as Bombelli's algebra for instance, bears two dates, what would a conscientious historian do in such a predicament? Write a paper on the discrepancy and submit it for publication to his favorite learned journal, thereby possibly starting a priority row. This again has happened more than once, usually over trivia.

Leaving Diophantus to dangle in the "misty mid region" of history, let us pass on to Hypatia. She is credited with the first recension of Diophantus' *Arithmetica*, and this is accepted by Heath and others as the ultimate source. Yet nobody is recorded as ever actually having seen it. Heath says that a certain manuscript, *dating in the thirteenth century*, "the most ancient and best of all texts, is evidently *a most faithful copy of* Hypatia's [lost recension]." How does this rate as evidence? On what is that "evidently" based? Could it possibly be extrasensory perception or table-tilting? By the thirteenth century Hypatia had been dead about 800 years, and nobody is on record as having seen the original of this "most faithful copy" of her work. If substantiated documentary evidence is demanded, where does this leave Hypatia's recension? But as even the most exacting and most meticulous of historians

184

believe that she did compose a recension of the *Arithmetica*, mathematicians must do the same or dismiss Hypatia's recension, long since vanished, as a thirteenth-century fable. Unless it is accepted (on faith) for what it is said to have been, the work of some later commentators is based finally on nothing.

These remarks are not intended to impugn the asserted historicity of any persons or their works. They are offered merely to suggest that in the application of "documentary evidence," there may sometimes be a double standard of credibility. What is sauce for Hypatia should be sauce for Diophantus, or the other way about, or perhaps neither.

The loss of Hypatia's recension was a historical misfortune, but not a mathematical disaster. Plenty of other commentaries, or parts of them, were found.

Anyone who may think the skeptical young chemist quoted earlier was too brash or too disrespectful or too cynical, should consult Neugebauer, pp. 171–2, footnote. This pungent footnote reports on modern historical research by authorities far better equipped than was Heath in his day. The current conclusion: "We must admit that Diophantus cannot be dated with any accuracy within 500 years."

A similar job is long overdue on Hypatia's recension.

2. Collapse

After Hypatia's death there was an exodus of students from Alexandria to Athens, where there had been a few mathematicians, none distinguished, connected mostly with Plato's Academy. With the influx from Alexandria, the Athenian school came to life for a few years, and mathematics was again cultivated. But the creative spirit had gone out of it, and after a fruitless 114 years it expired. The second Alexandrian school

185

survived listlessly for about another century. The mathematical output was commentary and criticism of the masterworks of the golden age. Exhausted by this feeble effort the effete school collapsed, and Alexandria passed from the stage of mathematics. Both the Athenian and the Alexandrian schools succumbed to the hostility of bigoted religious intolerance. Athens was disposed of by a decree of the Christian Emperor Justinian I (483–565) forbidding the study of all "heathen learning" in Athens. Mathematics came under the ban, because Christians had made no contributions to the subject. Plato's Academy was closed (529). That date is a convenient marker for the beginning of the Dark Ages. The Academy had lasted almost a thousand years. The designation "Dark Ages" is sometimes applied to the earlier part of the Middle Ages—from the fall of the Western Roman Empire (476) to the revival of learning in Christian Europe, or sometimes to the entire period 400–1400. (I am aware that the term *dark* is objectionable to some scholars today, who deny that there ever was a "dark" period. They may be right. But nobody has yet claimed that the so-called Dark Ages were a blaze of brilliance in the field of mathematics.)

Justinian, with headquarters in Byzantium, governed the Eastern Roman Empire, most ably assisted by his empress, Theodora. *Theodora*, incidentally means "(female) gift of God," and Justinian's pet name for her was "my Gift." In gratitude to the Donor, no doubt, Justinian suppressed all "heathen learning."

Both Justinian and Theodora started quite low down, and when they reached the top, turned extremely pious, which was just too bad for "heathen leaning." Between them this remarkable pair acted out the greatest inspirational success story ever told. It might be titled "From Peasant to Emperor and From

Gutter to Throne." Either of these alone would be spectacular enough. Both successes befalling the same couple, a man and his wife, are almost miraculous. The lucky man was Justinian, famed for his codification of Roman law; the luckier woman was his empress, Theodora—famed for infamy.[3]

Justinian was a self-educated Macedonian peasant. But he had an influential uncle at the Byzantine court, who helped him up. A gifted man and an incessant worker, Justinian with his uncle's help rose rapidly. Theodora had no friends when, as a destitute young girl, she got a job in the Circus as a comic. Her childish antics won the applause and support of one of the political factions of the moment. They made a pet of her. An

---

[3] If anyone is interested in knowing what the Empress Theodora really was (according to Procopius), here it is: a sadistic nymphomaniac pervert, expert in all the permulations and combinations of fornication. This puts it plainly, but not too plainly. It would be amusing to know exactly which of her wiles she used to trap the morbidly pious and unworldly Justinian. In spite of his vast learning and indefatigable industry he seems to have been something of an innocent.

It was a time of desperate troubles. Though hot wars raged to save the empire from complete disaster, Justinian kept cool and continued his studies unruffled, confident that a man who was not an abject physical coward like himself would come forward. In the meantime, while waiting for the man of courage, Justinian played safe. His patience paid off and he did not have to venture near the danger zone. If a fight was in prospect three hundred miles away, he prudently advised a withdrawal to four hundred. His deputy in combat was Belisarius, who almost single-handed temporarily won back the Roman Empire from the Goths, Vandals and other "barbarians."

It seems as if Justinian in his pious study of the Scriptures must have pondered deeply the history of King David, Bathsheba, and Uriah. David saw to it that Uriah, Bathsheba's lawful husband, was accorded a hot spot in the front line. But that was for reasons other than the yellow cowardice which kept Justinian far behind the front. Even Theodora occasionally doubted the masculinity of the lord and emperor. But she had sense enough not to kill the overstuffed goose who, against all natural law, laid the golden eggs. So they lived happily together and between them extinguished "heathen learning." The courageous Belisarius, Justinian's stooge in military matters, was an extremely able general, but otherwise quite a sap dominated and hoodwinked by his wife and other female appendages. Voltaire dismissed him contemptuously as "that uxorious cuckold."

unprofitable early romance took Theodora and her man to Alexandria. The city was dull after the rip-roaring Circus, and the young Theodora left her sweetheart and skipped out. Living the life of a starved alley cat, she made her way back to Byzantium. There, after much fortune, good and bad, she captured the sober and studious, not to say stupid, Justinian. They made a perfect team. Theodora did not desert her "sisters in sin" when she became empress, but did all she could to see that they were at least decently fed. Justinian paid the bills out of the public purse. It is doubtless regrettable that Theodora, like some of the other ladies who have influenced science and learning, if only by suppressing both, would not have been acceptable to the Puritans. But, to parody Shakespear, God made them, therefore let them pass for women.

Anyone wishing more information about Justinian and his Gift will find some in the *Secret History* or *Arcana* of Procopius (born in the late fifth century). As official historian, Procopius, in his *History of the Wars*, was all appreciative respect for Justinian and Theodora; in his *Arcana* he told the truth as he saw it and blasted them both. He should have titled his *Secret History* "The Toady's Revenge." It is a masterpiece of long-hoarded contempt, venom and truth. Of course it was not published during the author's life-time. The *Arcana* has been advertised to adolescents as a juicy gobbet of hot pornography. It isn't; there is only one episode in the entire book that might amuse an American boy or girl over fifteen, and that one is not worth hunting for. As Oscar Wilde observed, it is cold mutton.

We know and admit that pagan learning was suppressed by Justinian. To keep some sort of balance, we must remember that the early Christians were sincerely interested in saving the souls of those they persecuted. Or so they said. On a less inhuman level, memories of the persecutions they had suffered

188

from the pagans of Rome may still have rankled a little. The ravening lions let loose on cowering Christians in the Amphitheater to enliven a boring Roman holiday were not entirely forgotten. The Byzantines had plenty of lions, but employed them for nothing bloodier than battles to the death with gangs of panthers in the Circus. The hedonistic Byzantines got their fun out of betting on the chariot races in their Hippodrome. As practiced by the Byzantines, chariot racing was an exciting and a dirty and dangerous sport. They also amused themselves by humiliating the pagan scholars, ridiculing them and closing their schools.

Not to leave the Byzantines with this unflattering picture, we may see what they were like in their more intellectual moments. Leaving aside their interminable and sometimes bitter debates about the nature of the Holy Trinity, I shall sample the offering of the genial Metrodorus in the early fourth century. He was responsible for the kinds of problems that occupied the Byzantines and the brighter Romans in the eight centuries before the Turks sacked Constantinople (1453). His contribution is recorded in the Greek, or Palatine, *Anthology*, Book 14, Epigrams 116–146, mostly a collection of simple numerical problems. Nearly all can be solved mentally except, possibly, by a mathematical imbecile; the Cattle Problem of Archimedes in its simpler form is the conspicuous exception. Epigram 126, already quoted, leads to an equation of the first degree in one unknown for the age at death, eighty-four, of Diophantus. An equally easy one typifies the way Metrodorus embroidered his problems to make them more attractive to the unmathematical eye, Epigram 118:

Myrto once picked apples and divided them among her friends; she gave the fifth part to Chyrsis, the fourth to Hero, the nineteenth to Psamthe, and the tenth to Cleopatra, but she presented the twentieth part to Parthenope, and gave only twelve to Evadne. Of the whole number a hundred and twenty fell to herself. How many did Myrto pick? [Answer, 380.]

Whatever else this may do, it preserves the beautiful Greek names of six girls. Other books of the *Anthology* allude to persons we have already met. For example Book 16, 325 is from a statue of Pythagoras:

The sculptor wished to portray not that Pythagoras who explained the versatile nature of numbers, but Pythagoras in discreet silence. Perhaps he has hidden within the statue the voice he could have rendered if he chose.

Book 7, 93 is on Pherecydes, who is said to have instructed Pythagoras:

The end of all wisdom is in me. If aught befall me, tell my Pythagoras that he is the first of all in the land of Hellas. In speaking this I do not lie.

Book 9, 400 is dedicated to Hypatia:

Revered Hypatia, ornament of learning, stainless star of wise teaching, when I see thee and thy discourse I worship thee, looking on the starry house of the Virgin;[4] for thy business is in heaven.

---

[4]The constellation Virgo. This and the other quotations are from the translation by W. R. Paton, London, 1917, reprinted 1925, 1948. Available in the Loeb Classical Library, Harvard University Press.

Finally there is one on Cleopatra, Book 9, 752:

I am Drunkenness, the work of a skilled hand, but I am carved on the sober stone amethyst. The stone is foreign to the work. But I am the sacred possession of Cleopatra: on the queen's hand even the drunken goddess should be sober.

When Justinian and Theodora closed the Athenian schools (including Plato's Academy), the ousted philosophers took refuge in Persia, where they were enthusiastically welcomed at the court and shown all the latest novelties of Persian passion. The philosophers were shocked. They had forgotten that they were descendants—intellectually, morally, and culturally—of the Ptolemaic Alexandrians, who thought nothing of a little polygamy between friends or of cozy incest among themselves. They had forgotten Platonic and Socratic love. But the Persians hadn't, and the prudish exiles from Athens got back home as fast as they could. Gibbon remarked that the philosphers' disapproval of the Persians' manners and customs was a most unphilosophical attitude for the philosophers to take.

The *Anthology* in its present form is the work of Maximus Planudes of Byzantium in the twelfth or thirteenth century, whom we shall meet shortly as a commentator on Diophantus. At this point chronology directs us to India and Arabia, but it will be simpler to finish briefly with the commentators and transmitters of the work of Diophantus, although some of them overlapped the Arabs or came later. The list is nowhere nearly complete, and I note only those few items in the work of a man relevant to our main purpose of getting from Diophantus down to Fermat.

Only six of the reputed thirteen books of Diophantus' *Arithmetica* are known. Hypatias's lost commentary on the first six books was the earliest in Greek; it *may* have extended to the

191

remaining books. The second Greek commentary was by the Greek Pachymeres (1242–1310) of Byzantium. He is a typical example of the inexhaustibly prolific scholar who writes many impressive works destined to be quickly forgotten. He wrote on philosophy, history and mathematics. Instead of giving a straight translation of the *Arithmetica*, he produced a paraphrase, of which only a fragment dealing with trivialities of definitions survives. Among his other learned works is a history of Byzantium in thirteen books. To relax himself he composed his autobiography in verse. On top of all this and more of the same, he held civil and ecclesiastical offices.

A third Greek to comment on the *Arithmetica* was that Maximum Planudes (1260?–1330) who straightened out the Greek *Anthology*. An honest and conscientious scholar, Planudes did not bowdlerize what he found. As an orthodox and practicing monk he must have been quite amused by certain flowers, in Book 12, of the provocative nosegay. Planudes was a scholar's scholar. The mass of his work is said by those who have seen it to be prodigious. He was a polymath; among other things he argued theology, sought out the finer pendantries of rhetoric, and wrote verse. He lived most of his life in Byzantium, but escaped for a spell as ambassador to Venice. His commentary on the first two books of the *Arithmetica* contained nothing new. It gave the definitions and statements of the thirty-nine propositions of Book 1, with detailed and superfluous elaborations; likewise for Book 2 and its thirty-five propositions. He also left much unpublished work. His life and his achievements fit well into the general decadence of Byzantium. We must now leave the happy Byzantines to frolic in peace undistracted by attempts to think.

Some readers may have missed at least an allusion to the mathematical contributions of the Romans. They made none worth remembering, nor in fact any at all. As for numbers they bequeathed us their childish "Roman numerals" and their clumsy scheme for writing numbers (for instance, IV, VI, IX, XI, XXIV, XXVI, meaning 4, 6, 9, 11, 24, 26), both of which nuisances, though laboriously reported in the histories of mathematics, are well worth forgetting. The Roman talent was for technology—durable military roads, architecture, building, viaducts, aqueducts and sewage disposal, in the last of which they excelled. They were also fairly good in war before they degenerated. The great Julius Caesar, for example, in his campaign against Gaul exterminated a million nearly helpless men, women and children, and enslaved that many more. This record, in passing, was only a first rough approximation to what the pundits of the United Nations called genocide in the trials of the "war criminals" after the Second World War. A new word may be needed for the atomic war. And there may not be another Cleopatra to make a ninny of the ultimate Caesar strutting to the victory parade through a shambles of shattered brick and erupting sewers. To immortalize his conquests, and himself, Caesar left posterity his *Commentaries*, a costive classic of dry Latin prose high up on any list of the world's dullest books, for generations the torment of bored teen-agers "taking Latin."

The Romans seem never to have heard that "All they that take the sword shall perish with the sword." The various brands of Goths, Vandals, and their kin remembered what they and other barbarians (non-Romans) had yielded to the Roman short sword, and in the fifth century got back some of their own. In 410 the last Roman garrison was withdrawn from Britain and recalled to Rome to bolster the crumbling defenses of the Western Empire. Gaul, North Africa and Spain broke off. In 476 the tottering remains finally fell, making a clatter

193

throughout Christian Europe like the racket of a drunken general falling down a long and steep flight of stairs. "When Rome falls, then ends the world." For classical conceit that statement would be hard to beat. Byzantium, capital of the Eastern Empire, still on its feet, had 977 years to go before the Turks finally knocked it over. Alexandria when Rome fell had 165 years left till the Moslems practically finished it. Neither Byzantium nor Alexandria of course ever wielded any such clumsy power as Rome had, so their collapse made less clatter than Rome's. Byzantium had recovered from its first sack (1204), when a gang of roistering Christian Crusaders took time out from their abortive mission of "redeeming" the Holy Sepulchre in Jerusalem, to pillage the rich city. They were seconded by the infidel Turks who, having thoroughly looted Byzantium (1453), became the progenitors of the Revival of Learning by scattering Greek classics and scholars broadcast over Christian Europe—not that the Turks gave a damn about learning; it was just an accident of war. The decisive kick they gave a lethargic culture was a thoughtless reflex action of military inconsequence.

3. Recovery

Though the devout Christian monks while working and praying kept some elementary and very crude arithmetic alive during the Middle Ages, partly to keep track of holy days, the blaze of European mathematics in the fifteenth and sixteenth centuries was not sparked by anything that originated in the monasteries, but by the infidel Moslems, Arabs and Persians. The Moslems in their turn had been indebted to the earlier Indians (Hindus), and they to the Greeks, including Diophantus. Both the Indians and the Greeks, as pointed out by Neugebauer, probably took off from the spread, both to the east and the west, of

in their spare time on a store of Greek learning, including mathematics they could never have absorbed through their heads, even by osmosis. But they had preserved it. Al-Mamum's commissioners were cordially welcomed in Byzantium. They brought back to the Caliph a rich haul of Greek learning. Mamum promptly set up a "stable" of scholars and translators to turn all this priceless Greek stuff into intelligible Arabic. The earlier translations were not very good; but experience brought skill, and some of the definitive sources of Greek mathematics are in Arabic. Let us not overlook our indebtedness in this decisive episode to the scientifically backward but happy Byzantines. Without their conservation of the Greek mathematical masterpieces there might not have been a significant Moslem mathematics, nor a mathematical revival and maturity in the Europe of the fifteenth, sixteenth and seventeenth centuries.

At this point I by-pass the historical chore of reporting on individual medieval Greeks, Byzantines and Arabs[6] (including the kind and unkind things that have been said about what

---

[6]I refrain from citing the few Arabs who might be chronologically considered here; none made any contribution worth noting to the theory of numbers. Historians who know Arabic say there are literally hundreds of Moslem mathematicians and astronomers whose work is still unknown to those who do not read Arabic. It may be so; but if it is, the following caution from M. Cantor (the mathematical historian, not to be confused with G. Cantor the mathematician), regarding the numerous unsung heros of Arabic mathematics is worth a passing mention: "Many names [of Moslem mathematicians in a certain German history of mathematics] are as dead as their books; let us take care that we do not resurrect them." Estimates of Arabic mathematics differ quite widely from uncritical enthusiasm to its sour opposite. Perhaps too disparaging is the application to Arabic mathematics ics of the famous critique of that rude bear, Samuel Johnson, on women preachers: "Sir, a woman preaching is like a dog's walking on his hind legs. It is not done well; but you are surprised to find it done at all." So say those who do not accept uncritical evaluations of Arabic mathematical genius.

these mathematicians did to get Diophantus down to Fermat), and I shall report very briefly on only one outstanding man. (For the rest, neglected here, I refer the reader to Heath's summary, available in almost any university or large public library.)

The one man to be noted is Xylander, whose name is the Grecized form of the German (Wilhelm) Holzmann (1532–1576). According to his autobiographical account, he was a public teacher of Aristotelian logic at Heidelberg, and was erudite to the nth degree. He had mastered the German and Italian works of his age on algebra, particularly "surds," and was quite set up about his mathematical ability. Then, in his voracious and gluttonous reading, he encountered Diophantus' *Arithmetica*. It stopped him. Here was mathematics. He realized that all he had previously digested was nothing compared to this. To recover his self-respect he decided to find a reliable manuscript of the *Arithmetica* and translate it into Latin. The Greek was crabbed, and like anyone making a first, and largely successful, attempt to transport difficult mathematics from one language to another, Xylander in his translation was not always happy. Nevertheless he did what was required, as will be noted in a moment.

Xylander was a most engaging personality and a museum specimen of that almost extinct bird, a genuine and sincerely modest mathematician. His account of his struggles to understand the *Arithmetica* is an unselfconscious classic of modesty. This intriguing mathematician and self-effacing scholar was responsible for finally getting the *Arithmetica* into the hands of Fermat. We shall see in the next chapter how the French Bachet finally put it where it rightfully belonged.

Bachet's treatment of Xylander, on whom he evidently had leaned heavily, was shabby and ungenerous to a degree; but let that go. It is all nearly forgotten anyhow.

8

The Last Euclidean:
Bachet (1581–1638)

Some of the men given fairly extended notices in the following chapters would have been forgotten as mathematicians or scientists long ago had it not been for their connections, sometimes distant, with Fermat. As concerns Fermat himself, but for a lucky and unpredictable accident he too might have been only one of a fading crowd, mentioned with pious respect and dismissed with equal piety in the standard histories of science and mathematics. In the decisive decades of the transition from ancient and medieval science to the modern (post-1600) physical sciences and mathematics, as initiated by Galileo, Descartes, Newton, and Leibniz, Fermat finds and takes his proper and dominant place. Fermat's extraordinary luck was his encounter with Bachet's Greek and Latin edition of Diophantus' *Arithmetica*. Until Fermat read and digested that book, much of his own mathematics had been brilliantly obsolete in the then fossilized tradition of the greater ancient Greeks. If Fermat was lucky in having Diophantus accidentally thrust at him, the current theory of numbers, with its numerous ramifications, was and is equally lucky in getting its first creative modern pioneer from that chance encounter. So the not-always modest Bachet merits more than a casual nod in passing.

The last considerable mathematician to write with the rigid formalism (but not in the form) of Euclid was Claude-Gaspar Bachet, Seigneur de Méziriac, 1581–1638, erudite classical scholar, poet, amateur mathematician and, most important, editor of the particular edition of the *Arithmetica* of Diophantus used and annotated by Fermat. Without Bachet's edition, Fermat might never have taken much interest in the theory of numbers. The generous margins of the pages tempted him to make notes, one of which was his famous Last Theorem. Bachet's own work in the theory of numbers also suggested problems to Fermat.

The publication in 1637 of Descartes' analytic geometry marked the end of the strict Euclidean tradition and the beginning of more flexible methods in all mathematics. This progress left Bachet far behind, though his edition of Diophantus started an epoch in the theory of numbers.

On both sides Claude-Gaspar Bachet came from scholarly and distinguished families. His father was judge of the appellate court of Bresse and a counselor to the Duke of Savoy. His paternal grandfather, Pierre Bachet (1510–1565), the Seigneur de Méziriac, was a king's counselor, an eminent jurist and an accomplished composer of Latin verses. The large number of magistrates and king's counselors in the sixteenth and seventeenth-century France is partly accounted for by the exorbitant taxes that had to be squeezed out of the peasants to support the royal family in extravagant and degenerating luxury, and to pay for almost incessant and profitless wars. Exceptional zeal in tax extortion was frequently rewarded by promotion into the ranks of the lesser nobility, many of whom might have afforded taxes but paid none. Pierre Bachet, born into the nobility, consolidated his position by marrying Françoise de Soria, a Portuguese lady whose father had been raised to the nobility for his services as chief physician to the Portuguese royal family. Françoise was sufficiently educated to be able to appreciate Pierre's Latin verses. Claude-Gaspar's father carried on the legal traditions of the family. His mother, Marie-Françoise de Chavannes, a daughter of the petty nobility, bore her husband six children. The French aristocrats were nothing if not prolific as long as the common people footed the bills. It was not a happy time for either the nobility or the peasantry. Chronic war had aggravated the poverty and filth of the provinces, breeding plagues that stopped only for lack of victims. Claude-Gaspar's father died (1586) of the plague. His mother remarried the following year, only to die of the plague shortly after, leaving Claude-Gaspar an orphan at the age of six. His stepparents, relatives Antoine Favre and his wife, treated him decently but somewhat impersonally, and saw that he got a sound classical and thoroughly unpractical education from the Jesuits. The stepfather was a man of note in his profession of the law, with a taste for versifying which he passed on to Claude-Gaspar, along with a firm ambition to "be somebody." Like many of his contemporaries, Claude-Gaspar almost ended up a nobody because of the handicap of knowing too much outdated lore.

On finishing his formal religious education the devoted young man endured a spell of teaching in a Jesuit College in Italy, although he did not join the Society of Jesus till later. When about twenty-one he was sent to Rome on a diplomatic mission to press the dubious claims of Anne d'Este, Duchess of Nemours, to succeed her brother to the duchy of Ferrara. While in Rome he amply occupied his spare time writing

Italian verse. Visits to Paris showed him how the best people of the day lived, grossly overfed like prize hogs at a country fair, and fussily beribboned like professional streetwalkers, male and female, in the midst of seething swarms of starving beggars and diseased cripples draped in rotting rags. Rather priggishly young Bachet turned his back on all the available fun. A social conscience had not yet been conceived in France, or elsewhere, much less born. It came to life some seventy years later in merciless hatred and uncompromising bloodshed when Brachet had long since ceased to bother his head about the diverting inequities of Paris and the waning servility of his servants.

Returning to his native countryside of Bourg-en-Bresse, Bachet retired to his pretentious estate with its quaint plumbing, to live a quiet, if rather dull, life of scholarly meditation and gentlemanly ease. Men of his class were not wilfully indifferent to the degraded suffering of the people whose labors made leisure and comfort possible for their social superiors. The squalid misery of the peasants was so far beneath the attention of the gentlemen that they were unaware of its existence. Bachet was like the rest. So long as his bailiffs squeezed out the last handful of grain the peasants owed their overlord—according to his reckoning, but not theirs—the aristocratic scholar was content. And so long as the innumerable courts of law dispensed justice according to the king's will, and a swarm of predatory tax collectors did their sworn duty loyally and efficiently, backed when necessary by brutal ruffians who called themselves, and were, soldiers of the king, the gentle country scholar asked no questions.

Except for occasional trips and stately participation in the stilted gaiety of the rural gentry, Bachet passed his days and most of his nights in his magnificent library, absorbing the classics of antiquity and the best of French and Italian science and literature. As a hobby he composed verse, mostly mediocre, in Latin, French, and Italian. Always a perfectionist, he was but seldom satisfied with his efforts even after years of fastidious polishing. Possibly the same quirk kept him a bachelor till he was forty, when he married Philiberte de Chabeau of the lesser nobility. He made up for lost time, however, as rapidly as nature would permit, by fathering seven children.

Like many sedentary scholars, Bachet suffered from various minor ailments, not all imaginary, and complained like a Babylonian about his health to anyone who would at least listen, if not sympathize. A real illness prevented him from going to Paris in 1634 to be welcomed into the recently founded Académie Française as one of its first members. Four years later he died at the age of fifty-eight, well beyond the normal span for men of his generation. Whatever faults he may have had were those of his times, for which he was not responsible. He was kind and gentle to those about him, and was liked and respected by all those of his own class who knew him. To console himself in his suffering from gout and rheumatism he translated a moralistic Italian work on tribulation, and to amuse his children wrote a life of Aesop. At the time of his death he was reputed to be the most learned man in France, especially in Hellenistic studies and antique classical mathematics. Diophantus could not have asked for a more competent editor.

Bachet's first and most widely known book was his delightful *Pleasant and delectable problems concerning numbers ... very useful for all kinds of curious persons who make use of arithmetic* (abbreviated title), publishing at Lyon in 1612, and dedicated to Cardinal du Perron.[1] Devoted to what we would call mathe-

---

[1] *Problèmes plaisants et délectables, qui se font par les nombres; Partie recuille de divers authers, et inventez de nouveau avec leur démonstration, par Claude-Gaspar*

matical recreations, the *Problèmes* was the first printed book of its kind. Its intriguing readability and its immediate popularity with "all sorts of curious persons" infuriated the sourer pedants. They stigmatized it as a frivolous and worthless treatment of a serious subject. Bachet, a scholar but no pedant, was presently to offer the outraged and envious critics something sober enough even for them in his Greek and Latin edition of Diophantus. Whatever his jealous critics might say, Bachet had undertaken the *Problèmes* soberly, partly to develop his expository skill and mathematical competence in preparation for the great *Diophantus*. The most celebrated predecessor of the *Problèmes* was the Greek (or Palatine) *Anthology*, a specimen from which we have seen in the "epigram" on Diophantus' age. Bachet published the Greek text of all forty-five arithmetical epigrams in the *Anthology*, with a Latin translation and solutions of the puzzles as a supplement to Book 5 of his *Diophantus*. He did not claim originality for the idea of a collection of problems like his. In fact mathematical recreations were popular in the Middle Ages, and even got into the schools to enliven the deadly teaching of arithmetic. (The two best known modern collections are those of that highly ingenious arithmetician Édouard Lucas [French edition 1891–96, vols. 1–4], and W. W. Rouse Ball, *Mathematical Recreations*, eleventh edition, 1939, revised by H. S. M. Coxeter, 1938, American edition, 1947. Both contain some of the old problems, most of which are too easy today, in addition to the newer ones, some of which are too hard for almost anyone but a trained mathematician.)

Bachet, Sr. de Méziriac. Très utiles pour toutes sortes de personnes curieuses, qui se servent d'Arithmétique. The second edition, 1624, "revised, corrected and increased by several propositions and problems by the same author" was dedicated to the Count of Tournon, and contained 284 pages against fewer in the first edition.

The *Problèmes* implicitly raised at least one question which is not likely to be answered for some time: how many magic squares of any given order are there? A magic square of *order* n is an arrangement of the n^2 numbers $1, 2, 3, 4, \ldots, n^2$, or some of them, in the form of a square such that the numbers in each row, in each column and in each of the diagonals add up to the same sum. (More general types of squares having the like constant-sum property have been investigated, but the foregoing is sufficient here.) As specimens of orders $3, 4$, we have

2	7	6		16	3	2	13
9	5	1		5	10	11	8
4	3	8		9	6	7	12
				4	15	14	1

The second is from Dürer's picture, "Melancholia," 1514 (middle numbers of bottom line). The square dates the picture.

There are many rules for the construction of magic squares, but nothing really usable for the problem of enumerating them all for any given order. The construction of magic squares attracted Fermat, who did not claim that his rules gave all. (The specimen Fermat gave shows that his rules were defective.) Writing to Mersenne in 1640, he justly criticized Bachet's work without significantly improving on it.

In his attack on the old problem of solving indeterminate equations of the first degree *in integers*, Bachet was completely successful. He gave the first general, European, nontentative method for finding the *integer* solutions of equations of the type

$$ax + by = c,$$

where, a, b, c are given integers. (The Indians with their pulverizer anticipated him, but without proof.) Lagrange paid Bachet's original method a deserved compliment: "It is very direct and very ingenious, and leaves nothing to be desired in the way of elegance and generality." The like could not have been said of the partial anticipations by the Indian and the Italian mathematicians, all of whom resorted to a certain amount of trial. Bachet's method is substantially that explained in simplified form in our textbooks.

Much of Bachet's published writing catered to the scholarly and religious taste of the age. It is extremely varied. In 1614–18 for example, he brought out in Latin an epistle that the Virgin Mary might have written to Jesus, a sheaf of Italian verses, short Latin poems both sacred and profane, translations of the Psalms, canticles, and a translation in verse of Ovid's *Epistolae* or *Heroides*—imaginary love letters from faithful ladies of the heroic age to their absent but supposedly still devoted husbands. All this and more of the like was his recreation. His main interest seems to have been mathematics in the synthetic manner of the Greeks.

Unfortunately for the permanence of all Bachet's laborious imitations of the ancients, the synthetic, Euclidean method was already antiquated when he wrote. The rapid evolution and spread of symbolism, especially after Vieta (1540–1603), had demonstrated its power in algebra, was a major factor in the rise of the analytic method and a main source of the incomparably greater creativeness of the new mathematics contrasted with the old. Bachet never mastered the use of symbols, either because it violated the purity of Euclid's frigid style and therefore was distasteful to him, or because he did not have the same type of mind as Descartes, whose free use of symbolism in algebra and geometry helped to revolutionize mathematics. Bachet's *Elements of Arithmetic* in thirteen books

(*Elementorum Arithmeticorum*, *Libri* XIII) modeled on Euclid's *Elements*, was a fossil of a taste that had died forever. There was hardly a symbol or an operational abbreviation in the entire book; except for an occasional plus or equals sign everything was fully written out in words, and the work remained in manuscript. Its great merit, rigorous proof, was obscured by the archaic style and soon forgotten. In some respects nothing so exact appeared till the *Disquisitiones Arithmeticae* (1801) of Gauss. For example, Euclid had proved the commutative law of multiplication, $ab = ba$, for products of two factors; Bachet proved it for three or more, which Euclid could not have done, and just missed a general statement of the principle of mathematical induction. Another reason for the indifference to Bachet's work was the contemporary neglect of the theory of numbers in favor of the new analytic geometry and the differential and integral calculus. Even Fermat's greatest work did not receive its due till Euler and Lagrange continued it in the eighteenth century long after Fermat was dead.

Bachet's mathematical and scholarly masterpiece, the work for which he is remembered, if only in connection with Fermat, is his Greek and Latin edition of the works of Diophantus. The first edition was published at Paris in 1621. Samuel Fermat (the great Fermat's son) reissued it at Toulouse in 1670 as the first of two volumes containing selected writings of his father. Bachet had written his own *Arithmeticorum* to prepare himself for editing the extant books of Diophantus' *Arithmetica* and the tract on polygonal numbers—*Diophanti Alexandrini Liber Sex, et de Numeris Multangulis Liber Unus*—"now first put out in Greek and Latin." His text was based ultimately on manuscripts in the Vatican Library. For help in some of the harder mathematical technicalities he relied on algebra as expounded by Bombelli (second half of the sixteenth century) and Vieta. To the tract on polygonal numbers he added an appendix in two "books," in which he included some of his own work—a general formula for the various kinds of polygonal numbers, among other items.[2]

Fermat's marginal notes on his copy of Bachet's *Diophantus* made mathematical history that today is of much more than antiquarian historical interest. What if Bachet had not edited the six books of the *Arithmetica?* Fermat of course could have used Xylander's Latin translation, provided a copy had come his way. Bachet's edition, printed in Paris, was readily available. History might have been different if Bachet had not edited Diophantus.

---

[2] He gave the equivalent of the formula

$$P_{r+s}^{(n)} = P_r^{(n)} + (n - 2)rs + P_s^{(n)},$$

where, for any integer m, $P_m^{(n)}$ denotes the mth polygonal number of n sides,

$$P_m^{(n)} = \tfrac{1}{2}m[2 + (n - 2)(m - 1)],$$

for $n = 3, 4, 5, 6, \ldots$. For the square numbers $n = 4$, and Bachet's formula gives $(r + s)^2 = r^2 + 2rs + s^2$.

9

Mathematician and Jurist—Fermat

The vital statistics of Fermat's life are uninteresting and unilluminating. Pierre Fermat was born in Beaumont de Lomage, the principal town of the department of Tarn and Garonne, on August 20, 1601. (This date has been disputed; the baptismal date August 20, 1601, is documented by church records.) The great Fermat's father, Dominique Fermat, was a merchant and second consul of the town of Beaumont; the name of his mother was Françoise Cazaneuve, and that seems to be about all that is known of her personally. I must interpolate here a note about Pierre Fermat's legal activities. First a lawyer (as we should say), he became a councilor at the Chamber of Requêtes (petitions) of the Parliament of Toulouse. His appointment as councilor dates May 14, 1631. Some days later he married Louise de Long, daughter of a councilor of the same Parliament and a *cousine* (companion) of his mother. Five children were born to this marriage, including two who became nuns. The only one of any mathematical interest was Clément-Samuel, a magistrate, who made a fragmentary attempt (1679) to edit his father's mathematical remains. Though incomplete, this work proved valuable.

Disappointingly little is known of Fermat's education and his progress as a scholar. He was educated in the "higher school" of the Cordeliers, a Franciscan order and, to judge by his profound knowledge of the classics, including the mathematics, of antiquity, he must have received a solid grounding in Latin and Greek. He also was proficient in modern languages. As we look over what he did, Fermat sometimes appears as another of those lucky men who were not crushed by the weight of their erudition. At first he was unduly respectful of the geometry of Apollonius and the spirals of Archimedes. These things were "noble [frozen] and nude [bare] and antique [antiquated]" when Fermat and some of his contemporaries slaved over them, but they belonged to the past, largely because Fermat and Descartes had renovated geometry and algebra. Special problems were more or less abandoned for considerations of extensive classes of problems united by bonds of a certain generality, for example Cartesian geometry.

Except for short trips connected with his legal profession, Fermat lived out his life at Toulouse; he died at Castres, where he had gone to try a case, January 12, 1665. But for a rugged constitution he might have been eliminated earlier. The Thirty Years' War (1618–1648), which broke out in Germany between the Calvinists and the Catholics, was a by-product of the religious Reformation. When the war ended, central Europe was a stinking, disease-riddled shambles ravaged by starvation. To keep some spark of life in their all but fleshless bodies, the famished humans in many sectors preyed upon their weaker fellows, who were incapable of protecting themselves. They devoured the fresher bodies of the fallen and pitched the sheerly inedible out on the middens and dunghills to rot with bloated horses in an intolerable stench. Only the realistic rats took advantage of this unexpected bounty. Waxing fat and insolent, they defied the humans and proceeded to invade such parts of Europe as they could reach by fairly easy stages through the filthy towns and filthier villages, spreading plague wherever they lingered to feast. As for this war, fought in the name of idealistic abstractions, everybody lost. Such wars for ideals—justice, faith, honor, patriotism, freedom, peace, duty, loyalty and so on—are usually especially vicious and inconclusive; this one left only unappeasable hatreds as its legacy to civilization.

Fermat was seventeen when the war broke out; when it ended in all but total ruin he was forty-seven. His greatest work, including his Last Theorem, was then a decade behind him. Though he continued to do mathematics and to practice law, his vigor was not what it had been in his prime. It is said (by Pascal among others) that moderate attacks, possibly two, of the plague had sapped his vitality; indeed Fermat intimated as much to Pascal.

We must return for a moment to a detail in Fermat's legal activities. He has been cited as a "master of *requêtes*" (petitions). Who were these men? As French scholars themselves seem to find it difficult to describe their multifarious duties and their involved history, I shall note very briefly only what applied to Fermat. It is all mixed up with the evolution of French medieval law. At first the *maîtres de requêtes*, theoretically at least, undertook to put the petitions of his subjects directly before the king. In the beginning there were but two receivers of petitions. Then the number increased to five, then to six, then to eight, where for a while it stopped. Then the purchase of lucrative public offices increased so rapidly that the bureaucratic lid blew off: there was a judicial section, a "Chamber of Enquiries, a Chamber of Requests, and a bicameral Chamber, all dominated by the 'Great (Grand) Chamber,'" where the king came to preside when he felt so inclined. This organization embraced the peers, including representatives of the Church and of the University of Paris. Only the most important cases got this far up the ladder of legal consideration. The main function of this chamber was reporting on inquiries or investigations that had originated at a humbler level. The "Masters of Requests of the Palace" examined petitions and handed up "letters of justice" which authorized the bringing of them before Parliament, or getting permission to plead through an attorney. They also protected, more or less inadequately, the rights of widows or officers of the king's household. Later they were allowed to issue letters transferring cases to other tribunals. All this elaborate machinery for the prosecution of justice one might think would have ensured the people a modicum of equity. It did nothing of the sort: privileges were distributed over a random mob of members of the nobility, officers of the crown, members of Parliament, public servants, and many others, resulting in wholesale graft and abuse.

When we come to Fermat the jurist (Chapter 14) we may imagine that he fitted willingly into this pattern of absolute loyalty to his king and his God with its accompanying disregard for the lowlier inhabitants of France. This might be hastily inferred from the crude evidence, so it is gratifying to record a conspicuous exception showing that Fermat was not wholly indifferent to the plight of his fellow men. All his life he made Beaumont, where he was born, his special concern. He softened when he could the royal decrees handed down from on high, and made it his concern to see that the peasants understood what was going on. As they naturally knew no Latin, he translated and explained to them what the decrees meant. He also saw that they were not robbed of their established folk customs.

Anyone privileged to live in a major earthquake belt will know that a fault is a crack in the earth's crust, sometimes extending down for hundreds of miles. Opposite sides of a fault slip occasionally, and when they do there is an earthquake. After the slip the countryside may be quite different topographically from what it was before; a straight road, for instance, intersecting the fault may be displaced into two roads, neither coinciding with the original. Fermat's mathematical life was like that. The fault which rearranged it occurred in his early thirties. Thereafter he became great—not merely a competent and erudite scholar, but a great mathematician. Until that accident in his thirties, Fermat was almost ignorant of Diophantine analysis and the theory of numbers. After the split he moved at once into the first rank of creative mathematicians, where he has remained ever since.

It must not be supposed that his work before the split was negligible or of but slight interest. Some of it would have made a lasting reputation for any man but Fermat. He was not alone in some part of it, as will appear shortly. But others given the time might have done piecemeal the things he took in his stride. Some instances follow.

Fermat's work on what are called the singular points of plane curves helped to open up the field of algebraic geometry. In 1636 Roberval proposed the problem of points of inflexion, which Fermat solved. The year 1638 was marked by an irreconcilable quarrel with Descartes over the matter of tangents, essentially a problem in the differential calculus. Fermat was right, for which Descartes never forgave him. Part of the trouble arose from Fermat's custom of exposing his mathematical discoveries with excessive conciseness. Again, in 1636, using Archimedean methods, he generalized Archimedes' spirals and investigated their geometry. The study of the simpler singularities of plane curves (points of inflexion, etc.) remained a major interest of Fermat for years. All his discoveries in this field were long since included as exercises in the old-fashioned texts

215

on the calculus. Their interest seems to have evaporated, even for students; there is so much of living mathematics to be done that there is no time for the appreciation of fossils. Not so long ago texts on the calculus were cluttered with these relics of the seventeenth century and others like them. But in Fermat's hands all these antiques led him to his forms of the differential calculus and analytic geometry, both major advances. In another direction he inaugurated the theory of maxima and minima, giving, for the simpler types of problems, the rules and procedures of our texts today. This later (but not in Fermat's hands) led to a major department of modern analysis, the calculus of variations. In his study of maxima and minima, Fermat arrived at his Principle of Least Time: a ray of light traversing a medium susceptible to reflection and refraction will go by the path taking the least time of all possible paths between two fixed points in the medium. (This, of course, needs amplification to make it exact.) It is interesting that Pappus of Alexandria (second half of the third century) was acquainted with a simple example of the general principle, sufficient for, say, a system of plane mirrors. Historically Fermat's principle is of interest because it provoked a violent attack from Descartes. To minimize his misunderstanding of the minimal problem and its solution, Descartes modified a statement here and there, then proceeded to deduce nonsense from the travesty. Fermat ignored Descartes' dishonesty, which only made Descartes angrier and stupider than ever. Fermat held his tongue, the philosophical Descartes erupted in letters. But Fermat won the debate because he was right and Descartes wrong.

In 1636 Fermat gave his personal account of the integral calculus. An interesting detail is his application of the polygonal numbers to the summation of certain series. As a last example of Fermat's overall mathematical power, his part in the creation of the theory of probabilities may be cited. In a short correspondence (1654) with Pascal, he and Pascal laid down the foundations, extremely simple, on which the theory of probability rests today. Pascal and Fermat proceeded by different paths toward the same goal. Some of Pascal's conclusions and calculations were wrong; all of Fermat's were right. The entire theory grew out of a question (duration of play) in games of chance. Fermat considered both independent and dependent probabilistics (i.e., "product" and "sum"). The basic algebra he developed is as in any elementary text on probability, or in the chapter devoted to it in college algebras today. In view of Pascal's mishaps in this matter his opinion of mathematics is revealing. Writing to Fermat he says: "To speak to you frankly about mathematics, I find it the highest exercise of the mind, but at the same time I recognize it as useless." Having mentioned algebra, I shall cite Fermat's method of elimination. Though frequently laborious, it is effective.

This list (incomplete) of Fermat's achievements outside the theory of numbers is enough to substantiate his claim to having been one of the major mathematicians of the seventeenth century. A topic that Fermat frequently refers to in his correspondence, magic squares, described here in Chapter 8, has been omitted entirely because it seems devoid of mathematical interest. (Not so for "Latin squares," which were suggested partly by magic squares; these are important in agricultural experiments.)

We must now follow Fermat across that decisive rupture in his mathematical life. Up till 1636 he had taken only a perfunctory interest in the theory of numbers. Then, ironically, his curiosity was captivated by a very simple problem in Diophantine analysis which he found difficult. Writing to Roberval in 1636, Fermat says, "Permit me to change the subject and ask you for a demonstration of this proposition which I frankly confess I have not been able to find, although I am assured it is true":—in modern language: If a, b are rational, and if

$$a^2 + b^2 = 2(a + b)x + x^2,$$

then both x and x^2 are irrational. Roberval reduced the problem to showing that $2(a^2 + b^2 + ab)$ is not a perfect square, which he disposed of by a simple argument on parities (evenness, oddness) of the numbers concerned. It seems incredible that this problem should have baffled even for a moment the master arithmetician who was shortly to be proving such gems as "a prime of the form $4n + 1$ is a sum of two square integers in one way only."

A slightly later problem is that of Fermat giving "a characteristic of the number seven: Apart from the obvious solution $x = y = z = 1$ the only solution in positive integers of the system

$$2y^2 - 1 = x, \qquad 2z^2 - 1 = x^2$$

is $x = 7$, $y = 2$, $z = 5$."

This was first proved in 1883 by A. Genocchi,[1] after Pepin, a fine arithmetician, had given it up. The proof was by means Fermat might have used.

With these initiations to Diophantine analysis we shall pass on to an account of some of the men who worked in the theory of numbers alongside of him, even though he maintained his freedom and independence. Fermat was not a collaborator.

---

[1] *Nouvelles Annales de Math.* III (1883), 306–310. See also Chapter 10.

The most influential of Fermat's contemporaries in the theory of numbers (and other fields of mathematics) was Mersenne. He is even more important than Bachet for Fermat's arithmetica. Not that he himself contributed much to the subject; he was a go-between for Fermat and others. Mersenne was so active that he merits an extended account.

10

The Catalyst:
Mersenne (1588–1648)

Through the medium of numerous journals published at regular intervals, particularly the "abstract journals," current advances in science and mathematics are available to anyone interested. At least an abstract of an article is published a month or two after the article in full has appeared. In the seventeenth century, communication of ideas was by letter, and was not much farther developed than the "posts" in the days of Xenophon. In the exchange of ideas, Mersenne was a kind of general post office for some of the leading mathematicians and men of science of his day. He himself is rapidly fading into oblivion, and is remembered mathematically for his bad guess on perfect numbers, to be reported later. Probably no other guess in the theory of numbers has occasioned so much laborious computation as Mersenne's erroneous assertion. Though finally shown up, he has bequeathed us, by implication, a problem on primes that is still open. Mersenne's unique historical importance was his gift for stirring up profitable controversies among his friends and correspondents. Contributing little himself, he was a catalyst speeding up the exchange of ideas between others. He and Fermat, always on good terms, were frequent correspondents.

Mersenne's life overlapped Galileo's (1564–1642) for fifty-four years, Fermat's (1601–1665) for forty-seven, and Descartes' (1596–1650) for forty-six. Galileo is usually considered the first effective practitioner of the mathematical-experimental method in the physical sciences which we still follow. Descartes' geometry of 1637 and Fermat's earlier work of 1629 in the same direction marked one definite beginning of modern mathematics. "Fermat's way of drawing tangents" to curves, according to Newton (1642–1727), suggested his own method of the differential calculus. The French, including Lagrange, claim Fermat as the real originator of the calculus; but as Fermat's pioneering work was far surpassed shortly after his death by Newton, there is no need to argue the point here. Fermat and Pascal (1623–1662) share the invention (1654) of the mathematical theory of probability, as now practiced, without serious competitors, and Fermat was alone in his founding of the theory of numbers as the subject is understood today. These names are enough to suggest that Mersenne lived in one of the critical periods of scientific history. His letters show that he realized what was taking place. He knew that his contemporaries were going far beyond the greatest of the ancients—Euclid, Archimedes, Apollonius, Diophantus. That some of the moderns of the seventeenth century, notably Fermat, had started where the ancients left off, and acknowledged their heavy indebtedness to the past, did not minimize the impact of a major revolution in scientific thought and method. Though Mersenne's part in the revolution was largely that of an alert and intelligent spectator, he had his own importance, especially for Fermat. So we must see who and what he was.

221

Marin Mersenne was born at Oise, Department of Maine, France, on September 8, 1588. After an elementary schooling at the Collège de Mans, he entered the then recently founded Jesuit College of La Flèche in January 1604, as one of the first pupils. Descartes, a frail boy, followed about four months later. Mersenne was only six, Descartes eight. The friendship formed between the two little boys lasted, though in later years it did not always run smoothly. Like some others of Mersenne's friends, Descartes objected to Mersenne's custom of publishing sections of his work without permission. But till Mersenne left la Flèche in 1609 to go to the Sorbonne in Paris, he and Descartes got on well enough. Mersenne was more profoundly affected than Descartes by the religious education at the Jesuit College. He lacked Descartes' deep vein of skepticism, and perhaps on that account lived a more tranquil though less productive intellectual life than his friend. If Descartes was constitutionally incapable of Mersenne's early works in theology, Mersenne could never have reduced any doubts he may have had to Descartes' abysmal "I think, therefore I am"—the nadir of skepticism.

In 1611 Mersenne left the Sorbonne and entered the cloister of the Minimite Order at the Place Royale, Paris. The following year, at the age of twenty-four, after serving a year's novitiate at Nigeon and Fablaines, he took holy orders. He spent the years 1615–18 as a professor at the College of the Minims at Nevers. In 1619, at the age of thirty-one, he was promoted to an administrative position which he did not want and for which he had but little aptitude. Disliking the work intensely, he resigned at the end of the year and returned to the cloister at the Place Royale. There he stayed for the rest of his life, except for the journeys to the Netherlands (1629–30) "to take the waters" for his health, the east of France (1939), Provence and Italy (1644–45),and to the west and southwest of

France. His journeys all had at least a secondary scientific purpose, either to meet and confer with men he had known only by correspondence, or to renew old friendships. Even the unfriendly gossip he brought back to Paris stimulated science. Everywhere he went he profited by discussions with intelligent men, including heretics and atheists, both of whom he detested.

Even for the seventeenth century, Mersenne's correspondence was voluminous. To oblige some scientific friend he would often write out a copy of what another had sent him. This altruistic service sometimes got him into disfavor with men who had intended their communications to be confidential. Descartes in particular had ample reason to complain, as his philosophical speculations sometimes came dangerously close to heresy and heresy was reproved in peculiarly unpleasant ways. Mersenne's own writing filled whatever time he had left from his correspondence.

Mersenne might make an interesting study for some psychoanalyst. Though not exactly a split personality, he had some of the marks—an almost fanatical regligiosity as a young man, overcorrected in later life by an equally aggressive rationalism. His earliest work was an elaborate and curious exegesis of the Book of Genesis. When he reached the sixth chapter something seems to have happened to his belief in the worth of what he was doing. Beyond that point nothing further appeared, and a mass of incomplete manuscript lay neglected and forgotten. Still in the earlier phase of his development, he released his dammed-up energy in tirades against atheists, of whom there were many at the beginning of the scientific revolution, heretics and unbelievers in the official religion, of whom there were more, and libertines, who appear to have composed the bulk of the civilized population. Nobody seems to have listened to him, especially when he began proclaiming that the new science was the only true faith and the one irrefutable evidence for the existence of his God. He was his own sole convert. His zeal for what he considered the truth resulted in excellent popularizations of science, respectfully admired by those who had no need of them and blithely ignored by others. An accomplished scholar, he published good accounts of the works of Euclid and Apollonius among the ancients, and of Snell (the index-of-refraction man in optics) and Kepler among the moderns. He similarly publicized several less eminent scientists with a firsthand knowledge of the works of his scientific contemporaries. He labored to make these pioneers of science widely known to all scientists. Some might not otherwise have been aware of what their contemporaries were discovering. Here he sometimes ran into trouble by incorporating the work of others into his own without permission. There was no question of plagiarism, as he always acknowledged the source of what he purloined and published.

As he got deeper and deeper into science he rebelled against his early indoctrination and rejected all metaphysics as scientifically and humanly worthless. The only science worth cultivating, he declared, and sincerely believed, was fact-finding. He was a choice combination of devout Catholic and logical positivist. Even in our own time there are such.

For mere bulk the output of Mersenne's letters and books almost passes belief. He readily convinced himself, if not others, that all his prodigious activity in the cause of science was motivated by his avowed mission of bringing heretics and atheists to his God. Whoever was merely told the truth (scientific facts as he saw them), he sincerely and naïvely believed, would be obliged to accept his personal version of the true religion. A practical obstacle to the success of Mersenne's rational program was the reluctance of the ecclesiastical and scholarly authorities, then as now, to accept some of the findings of science as truth. When he had to choose between ecclesiastical dogma and scientific fact as he saw it, Mersenne chose fact. He studied and accepted Galileo's revolutionary non-Aristotelian mechanics, and six years after Galileo had been condemned to silence by the Inquisition, he edited and published (1639) two of Galileo's works. From Galileo he learned a great deal more than facts of scientific importance. In spite of his professed distaste for all metaphysics, he could not help absorbing and publicizing a metaphysics of a new kind from his study of Galileo. This was no less than the scientific method as we understand it today, of which Mersenne gave the first (1634) clear statement. At the same time he insisted that the scientific method as practiced by his contemporaries made a sharp break with the ancients, whose masterpieces he knew at first hand and still admired. The moderns, he rightly insisted, by their new method, entered a vaster universe than any the ancients could ever have imagined.

Mersenne's insistence on experiment as the only profitable approach to nature caused a break between him and the philosophical Descartes. The breach widened when, in 1638, Mersenne provoked a controversy among his mathematical and scientific friends, Fermat, Descartes and Roberval, by showing each of them confidential letters from the others. While ethically indefensible, this betrayal of confidence greatly benefited science. Failing to profit by this unpleasant experience, Descartes two years later sent Mersenne the first part of his masterpiece, the *Discourse on Method*, with strict instructions that it was to be shown for criticism and suggestions only to a few trustworthy friends. Mersenne was so enthusiastic about the *Discourse* that he publicized it widely, even sharing it with Descartes' enemies. Once again science profited; the end, as with the Jesuits at La Flèche, had justified the means.

225

Mersenne's own scientific work contained at least some shrewd guesses. He asserted that it is impossible to square the circle, although a proof of this fact was two centuries or more beyond the mathematics of his time. In physics he stated that sound is essentially "a mode of motion," as it was to be phrased some two centuries later, and he was partly responsible for the decisive French experiments of Pascal, Petit and others (1648) on atmospheric pressure. The experiments were not his own idea; on his visit (1644) to Italy he had learned of the barometer from Torricelli. In the same year he published his *Cogitata Physico-Mathematica*, recording many observations and experiments and, incidentally, the famous assertion (notorious and discredited) about perfect numbers (mentioned before) for which he is chiefly remembered.

Mersenne's experience as an intermediary for the exchange of ideas between scientists and mathematicians had convinced him that there should be some organized society of men with scientific interests to do professionally what he had attempted as an amateur. Perhaps he realized that something more impersonal than the violent and irreconcilable quarrels which he had provoked between Descartes and Roberval over a geometrical problem would be for the good of science. His effort to bring scientists together resulted in a discussion club presided over by the brothers Dupuy at their house in Paris. Among those participating were the Pascals, father and son, and Carcavi, a judge, all friends of Fermat. Descartes when in Paris and Huygens (1629–1695) also took part. Fermat never visited Paris. The members of the club agreed that a scientific academy, recognized and subsidized by the French government, was what was needed, and in 1635 they organized the short-lived *Academia Parisiensis* (Paris Academy). This first attempt was not backed by the French government, and in 1666 was superseded by the Paris Academy of Sciences, created by Colbert

could not solve. A scrupulously honest man, Fermat took the accusation as an affront to his personal integrity. The fight was on. Wallis swung a clumsy bludgeon of assumed superiority; Fermat pricked at him with the lean rapier of Gallic wit and formal French politeness whetted by irony and contempt. (An example of Wallis' obtuseness in arithmetica, with Fermat's acid comment on it, will be given as the tailpiece to this notice of Wallis.)

Ineptitude in the theory of numbers does not necessarily imply incompetence in other parts of mathematics, and it does not in the case of Wallis. He was one of those youthful prodigies who never quite pay off. But for the mishap of living in the age of Newton, Leibniz, Descartes, and Fermat, Wallis might have been a leading mathematician of the seventeenth century. If he had had the sense to let Fermat alone, his reputation would be higher than it is.

John Wallis came of a prosperous family, and was himself in comfortable circumstances till his declining years. His people could afford to send him to good schools, then to Felsted, where he progressed so rapidly that by the age of fourteen he was ready for Cambridge, and at sixteen matriculated (1632) at Emmanuel College. Mathematics, largely self-taught, was his diversion. His primary interests were languages, read and spoken—Latin, Greek, Hebrew, French—and a pedantic logic. He also was proficient as a student in anatomy and music. This does not exhaust the tally of his accomplishments. His gift for mental arithmetic and his memory for numbers were phenomenal. He could extract square and cube roots to absurd lengths in his head (one square root went to fifty-three digits), and recall all the digits in the answers a day or two later.

His most spectacular gift was for deciphering coded messages, and he has a claim to being the first scientific cryptanalyst. (He was jealous of his reputation and made an enemy of

Leibniz by refusing to divulge his methods.) On graduating B.A. and M.A. (1637, 1640) he took holy orders and moved to London, where he became tutor to the sons of distinguished families. His fame was made when, one day at dinner, he was handed a cipher message concerning the capture of Chichester. He cracked it in two hours. This brought him to the attention of the military and the government. Years later William the Third appointed him official decipherer. The ciphers he solved now were much harder than his initial success; one took him ten weeks, another three months. For these feats, important to his stingy employers, he was lavishly rewarded in praise, which is more economical than money. Near the close of his life he complained to Pepys about the consistently shabby treatment he had received, saying he had borne up well enough under it till he was past eighty (he died in 1703, aged eighty-six), but now that he was an old man his sight, hearing, and strength were not as they had been in his younger days. "Put not your trust in princes"—nor in politicians, unless you are a fool.

Except for his naïve faith in his patrons, who accepted everything offered and gave nothing of value in return, Wallis was shrewd and materially successful. On the death of his mother (1643), he inherited her estate and enough money to enable him to do as he pleased. He chose ecclesiastical politics at a time when disputes over points of dogma were bitter and dangerous. It cannot be said that Wallis followed the straight and narrow path. In fact his progress between Church and state, and between the two religious sects of the time, was extremely devious. He testified against Laud, the Archbishop of Canterbury, and was partly responsible for the loss of that stiff-necked prelate's head to the ax before a large and enthusiastic audience on Tower Hill, London. That Laud was convicted on illegal evidence did not disturb Wallis unduly. He kept his own head. For his distinguished services to the crown he was appointed Savilian Professor of Geometry in the University of Oxford and keeper of the University Archives. His next steps up were a royal chaplainship, a part in the revision of the English prayer book, and the degree of Doctor of Divinity. All of these distinctions were no doubt adequately earned.

Wallis did not really wake up mathematically till he was thirty-one. He stopped being a dilettante and attempted to put some sound reasoning into the untenable theory of indivisibles, an early seventeenth-century precursor of the calculus. Though unsuccessful he found, more or less empirically, many interesting things that later mathematicians were to validate, more or less. His *Analysis Infinitorum* (*Analysis of Infinities*) delighted Newton as an undergraduate at Cambridge and inspired him to create his own calculus. A difficulty in the early days of the calculus was to find the derivative dx^n/dx of x^n when n is not restricted to positive integer values. What was needed was the binomial theorem for such unrestricted values. Wallis had supplied this and Newton exploited it. Wallis also worked successfully on the geometry of conics, mostly in the synthetic manner of the Greeks—already superseded by the analytic method of Descartes when he wrote. In geometry he discussed the cycloid, solving two problems proposed somewhat dishonestly by Pascal. He was less successful with the arithmetical problems (reported presently) sent to him by Fermat; in fact he failed stupidly and completely. How was it possible for a highly gifted mathematician like Wallis to go astray as he did in the mere understanding of what an arithmetical question was about? Perhaps the psychologists can tell us. If not, there is the simple mathematical fact that Wallis failed to distinguish between algebraic divisibility and arithmetical divisibility. This has been a common and disastrous blunder for (literally)

233

centuries. Wallis had many zealous companions in misfortune, including the great Cauchy.

His inordinate vanity and his waspish temper (except when presiding at divine service) got him into irreconcilable quarrels with all the mathematicians who tried to set him right when he was stubbornly wrong, including Pascal, Schooten, Leibniz, Descartes and others. Like a paranoiac in a tantrum, he abused and insulted some of those he disliked, especially the French, none of whom he liked or respected. When the supply of mathematicians ran out he favored philosophers and fellow theologians with his attentions. In a controversy with the philosopher and political theorist Thomas Hobbes (1588–1679), the *Leviathan* man, Wallis saluted Hobbes for his "geometrical imbecility." Hobbes might have returned the salute by mentioning Wallis' "arithmetical idiocy," had he been acquainted with it.

This is a sufficient sample of Wallis' social graces. He was a fair specimen of the irascible intellectuals of his time, brawling over a word or a plus or minus sign and challenging one another to ridiculous duels with paper swords. They did not always fight fairly. One ludicrous duel between Wallis and Huygens is classic. Huygens had made an astronomical discovery which he wished to conceal for a time. So he wrapped it up in a cryptogram and sent it to the Royal Society. Wallis replied with a phony anagram which could be read in many ways, enabling him to claim priority for his English friends. Huygens was taken in. He dropped his claim. When he realized that he had been duped he was quite sore. Wallis was not always as slick as that in his scientific dealings. He was one of the first members of the Royal Society and contributed much of his own work to its meetings.

I have reserved for a later chapter comment on the *Commercium Epistolicum*, etc., published by Wallis at Oxford in

1658. This was a correspondence in which Lord Brouncker, Sir Kenelm Digby, Frénicle, and Schooten participated. Wallis had published his lectures in 1657.

A single specimen will suffice to show how obtuse a fine mathematician like Wallis could be when he really put his mind to a simple problem in the theory of numbers he did not take the pains to understand.

In a letter of 1657 to Kenelm Digby, Wallis states that the complete positive integer solutions of the following equations are as indicated:

(1) $\qquad X^2 + 12 = Y^4,$ $\qquad X = 2,$ $\qquad Y = 2.$

(2) $\qquad X^4 + 9 = Y^2,$ $\qquad X = 2,$ $\qquad Y = 5$

(3) $\qquad X^3 - Y^3 = 20,$ $\qquad$ None

(4) $\qquad X^3 - Y^3 = 19,$ $\qquad X = 3,$ $\qquad Y = 2.$

These are reproduced (verbally) in Fermat's *Oeuvres*, III, 438, and were inspired by Fermat's famous assertions (long since proved) that the stated positive integer solutions of (5), (6) are *complete*.

(5) $\qquad X^2 + 2 = Y^3,$ $\qquad X = 5,$ $\qquad Y = 3.$

(6) $\qquad X^2 + 4 = Y^3,$ $\qquad X = 11,$ $\qquad Y = 5.$

Wallis seems to have been annoyed by Fermat's custom of issuing statements without proof, especially when the statements were proposed as challenges to the English mathematicians. Of (5), (6) he says, "Whether, true or false, I do not much care, as I do not see what great consequence would follow." That is one way of disposing of a challenge. He

235

continues, "Such negative determinations [as (5), (6)] are very frequent and are familiar to us [the English mathematicians]. [Fermat's (5), (6)] put forward nothing better or stronger than if I were to say"–among others of the same kind—(1)–(4). He concludes:

(7) *"It is easy to imagine innumerable negative determinations of the kind."*

Comment by Fermat: "I am always surprised that M. Wallis scorns what he doesn't know."

Apparently there is no reference to Wallis' remarks in Dickson's *History of the Theory of Numbers*, possibly because Wallis completely missed the point of Fermat's challenges. Actually, as is obvious, (1)–(4) are essentially trivial. If each of (1)–(4) is written as a polynomial $P(X, Y) = n$ (a constant integer) P is factorable in the domain of rational integers for (1)–(4), while for (5), (6) it is not. In each of (1)–(4) P is so factorable into a product of only two (irreducible) factors Q, R, and it is sufficient to test the compatability of $Q = a$, $R = b$, where (a, b) runs through all pairs of divisors of n such that $n = ab$.

For example, from (1),

$$(Y^2 + X)(Y^2 - X) = 12 = ab:$$

$$Y^2 + X = a, \qquad Y^2 - x = b;$$

$$2Y^2 = a + b, \qquad 2X = a - b;$$

$$(a, b) = (1, 12), (2, 6), (3, 4), (4, 3), (6, 2), (12, 1)$$

and since $2Y^2$ is positive these exhaust the possibilities. For positive X the only solution is

$$a = 6, \quad b = 2; \quad X = 2, \quad Y^2 = 4.$$

It would be interesting to know whether Wallis got (1)–(4) in this obvious way, which seems not too probable in view of his remarks about (5)–(6), or whether he was guessing from empirical evidence.

I regret that there is not space to report on the work of the Scotch James Gregory (1638–1675), a contemporary of Fermat for several years. He attempted, illusorily, to settle some of Fermat's problems. His genius, like that of most of his contemporaries, was for the continuous, not the discrete. But unlike Wallis, he did recognize the king of beasts from his claw where he saw it.[1]

2. Digby (1603–1665)

Sir Kenelm Digby is of interest to us chiefly because he is the only recorded reporter of an unpleasant incident (noted in Chapter 14 here) in Fermat's long and honorable career as a judge.

Though he was one of the first members (1663) of the Royal Society of London, Digby was an unscrupulous and versatile quack with no claim to scientific or scholarly distinction in anything from literature to chemistry. But as a prevaricator and a religious turncoat he was easily in the first rank, and as a boastful clown he was without peer. These talents should have

---

[1]See James Gregory, *Tercentenary Memorial Volume*, ed. H. W. Turnbull (1939), pp. 399–434.

237

assured him an interesting if rather disreputable career, and they did. Equally interesting was his part as an attempted intermediary between Wallis and Fermat, in which he exposed himself with reckless bravery and fatuous stupidity to the Frenchman's polished contempt.

Kenelm Digby's father, Sir Everard Digby (1578–1606), born a Protestant, turned Catholic in 1599. He inherited a large estate which he did not live long enough to mismanage thoroughly. His practical sense was nil. He was equally inept in his shifty politics. The date November 5, 1605, will be familiar to any who remember their English history as that of the abortive gunpowder plot of Guy Fawkes to blow up the king and Parliament. The urchins of London, if nobody else, still remember it with their penny-begging chant, "The saime old gaime, the saime old gaime, we carry on the saime old gaime." The plot fizzled and the conspirators, including the idealistic Sir Everard, were apprehended. After "severe torture" Sir Everard was hanged, drawn and quartered, leaving his son fatherless at the age of three. Kenelm, of course, was far too young to learn from his father's example the folly of meddling in politics and religion.

Though reared a Catholic, young Kenelm early turned Protestant, only to turn Catholic again. His education was scrappy and scattered. Six months in Madrid was followed on his return to England by two years at Oxford which he left (age seventeen) without graduating. At Madrid he had fallen in with Prince Charles, whom he trailed back to England. For his loyalty, such as it was, to the prince, he was knighted and rewarded with a roving commission to Italy. Congenitally incapable of doing anything in a straightforward and open way, he secretly carried on (age twenty-two) a bizarre courtship of his childhood playmate, "that celebrated beautie and courtezan," Venetia Stanley, whom he finally married. His next venture was a foray with two privateers to attack a French and Venetian squadron off Scanderoon. It was not much of a fight, but the squadron lost.

The death of his wife (1633) sobered him up for a spell. He entered Gresham College,[2] where for two years he piddled with chemistry and exasperated the professors with idiotic arguments. He was in one of his Protestant moods at the time. It cost him only a slight effort to lie presently to Archbishop Laud that he had been reconverted. Or was it a lie? Perhaps only God and Digby knew. Anyhow, in the Rebellion, he played both ends against the middle but always with a slight advantage in favor of his Catholic friends. Parliament finally caught up with him (1642–43) and relieved him of his estate. Digby was not dismayed. Somehow he raised the wind and got to Rome for two years (1645–47), where he enjoyed himself hugely by annoying the Pope. By expert bootlicking he wangled three visits to England (1641–51–54), the last for two years. Even the hardheaded Oliver Cromwell was suckered in by Digby's toadying, as also was the Royal Society of London. It is evident that Digby must have "had something" to fool so many of the people all of the time. But he never fooled Fermat.

His most brilliant feat of selling himself to any who enjoyed being swindled was his "Powder of Sympathy." This was a quack medicine, of which the miraculous secret had been confided to Digby in his youth by "a Carmelite friar," who may have existed only in Digby's wormy imagination. It could heal wounds without contact; a garter with dried bloodstains would do, though far from the wound. Digby once dusted the powder on such a garter belonging to a friend. The cure was complete.

---

[2]This college figures frequently in the history of mathematics. Sylvester, for example, applied for a professorship there, but was turned down because the resident professors were stupid.

Where have we heard anything like that recently? It would be impolite to say. We are farther advanced in "science" than Digby; we do not need powder or any other material aid; sympathy alone is sufficient.

The frequent claim by people who might know better that Digby understood something about the mathematics and science of his time is absurd. But he did irritate Fermat with his superficial ignorance, and that is fame enough for any mathematical ignoramus. His touted versatility in the humanities need not concern us here, nor need his press-agenting of Sir Thomas Browne, the *Religio Medici* man, be taken too seriously. He may not have believed what he said.

3. Brouncker (1620?–1684)

Viscount William Brouncker appears frequently in the correspondence of mathematicians in relation to Fermat. He was a man of some skill in the theory of numbers and was interested in Fermat's challenges to the mathematicians of England and the Continent.

William Brouncker's title was somewhat of a sham. His people were of the Irish nobility only by right of cash purchase. Pepys said that William's father bought his way into a peerage, paying £1200 for the title, and adds that the new nobleman was so broke after the purchase that he lacked a shilling to buy his supper. This sounds like the kind of story an Englishman would tell of an Irish nobleman, especially if it were untrue. Whatever the fact of the matter, young Lord Brouncker was well enough off to earn his degree in medicine at Oxford. His interests were varied, including linguistics and an unsuccessful attempt to reform the musical scale. In mathematics, among other items, he exploited continued fractions (he did not invent them), rectified the parabola and the cycloid, and obtained an expression for $\pi/4$ as an infinite continued fraction. In answer to one of Fermat's challenges, to himself and Wallis, he rediscovered the Hindu "solution"

$$x = \frac{2t}{t^2 - n}, \qquad y = \frac{t^2 + n}{t^2 - n}$$

of the Pellian Equation $nx^2 + 1 = y^2$, provided (as noted before) one solution could be guessed. Though this did not answer Fermat's challenge it was closer to the mark than Wallis' singularly stupid "solution" $x = 0, y = 1$. Fermat had asked for integer solutions.

Brouncker was very active and equally civilized in public affairs. He was a charter member of the Royal Society, incorporated under royal charter 1662–63, and was its first president till 1667, when he resigned; he was President of Gresham College for three years, 1664–67. Of all his honors he prized most the respect which Fermat had for his mathematical ability.

4. Frénicle (1605–1675)

Bernard Frénicle de Bessy, born at Paris, figures frequently in the correspondence of Fermat. His high skill in what many be called the inductive theory of numbers is frequently baffling, as he obtained correct results and only seldom gave a hint of how he had found them. Even Fermat was occasionally puzzled. (More on this will appear when we come to letters in the *Commercium Epistolicum* published by Wallis.) His methods apparently were empirical and elementary. In 1660 (age fifty-five) he abandoned mathematics and turned first to theology, then to saving his soul. On a mundane level, he was what we would call a civil servant. Except for his secretiveness about his methods he seems to have been a normal human being.

5. Carcavi (?–1684)

Pierre Carcavi, born at Lyon, is of unique importance in the history of Fermat's career. It was he who induced Fermat to state what he considered his most important discoveries in the theory of numbers when Fermat foresaw his approaching death. Only the statements without proofs were given, and to this extent Carcavi became Fermat's scientific executor.

Carcavi and Fermat were friends and colleagues in the Parliament of Toulouse till Carcavi moved to Paris and bought an office as a councilor in the Supreme Council. There was nothing objectionable about such a transaction in Carcavi's day. Carcavi was also a friend of both Pascal and Descartes till the latter, with his usual touchiness, provoked a row over nothing of consequence. On resigning from the Supreme Council, Carcavi made a high reputation for himself in bibliography. Colbert recognized his ability and entrusted his collection to him. His first job was to arrange and copy in 536 volumes the memoirs of Mazarin, which he accomplished in five years. (We shall meet Mazarin and Colbert later in unpleasant connections.) As a reward for this diligence, Colbert in 1663 appointed Carcavi custodian of the king's library. On the creation of the French Academy of Sciences, Carcavi became one of the leading members in the section of mathematics. He retired after the death of Colbert (1683) and died in 1684.

Fermat could hardly have asked for a better mathematical executor than Carcavi. It is one of the ironies of history that Fermat's page or two salvaged by Carcavi are probably better known today throughout the civilized world than the 536 volumes of Mazarin.

6. Huygens (1629–1695)

It may seem strange that Christian Huygens be included here. He is mentioned for two reasons: he figures directly or indirectly in the correspondence of Fermat; he marks the sharp break in interest between the theory of numbers and other, more popular, departments of mathematics, both pure and applied—horology, astronomy, lenses, the theory of light, etc. His name survives in the history and practice of physics. In actual arithmetica he made (so far as I can find) only one definite contribution to the theory of numbers, a solution of the Pellian Equation. But like all before Lagrange he did not prove that his algorithm (continued fractions) would always produce a solution.

Huygens and Fermat had the highest regard for each other and frequently expressed it in letters. If it were not obvious that both were first-rank men, we might be tempted to dismiss their friendship as just another mutual admiration society. But it wasn't.

7. Roberval (1602–1675)

G. P. Roberval's decisive part in sending Fermat on his way into the jungles of Diophantine analysis has been noted. For the rest, he wrote on the cycloid, the spirals of Archimedes, indivisibles (an early and untenable form of the calculus), and the elementary mechanics of his epoch. His contributions were on the whole competent and dull. Personally, he was somewhat of a snob and a fake: to ennoble himself with the coveted title which he lacked, he prefixed a "de" before the "Roberval"

243

(the name of his birthplace), and became, to himself, the Seigneur de Roberval. Significantly, he was one of the earliest members of the French Academy of Sciences, and by his peppery comments on the efforts of his fellow Academicians kept some of them alert.

8. Pell (1610–1685)

John Pell is a classic instance of a man who was neither born great nor achieved greatness; he had such notoriety as accompanies mathematics thrust upon him after he was dead. Though his name occurs frequently in histories of mathematics, especially in the theory of numbers, Pell mathematically was a nonentity, and humanly an egregious fraud. It is long past time that his name be dropped from the textbooks. But there it is, and there it will stay till mathematical historians take account of some plain facts in the technical records of mathematics.

Pell came of a respectable family in Sussex, England, and a passable education at a free school. Neither then nor in his mature life did he show the slightest sign of genius. By almost stupid diligence he earned his B.A. degree (1623) and his M.A. (1630); he was fairly competent in Greek and Latin. It is not clear how these modest attainments got him a professorship at the University of Amsterdam, where he lectured on, of all things, Diophantine analysis, about which he knew practically nothing. Next, the Prince of Orange, who knew even less than Pell about mathematics, prevailed upon the alleged mathematician to migrate to the new college at Breda. Pell's mathematical stock now rose rapidly: Oliver Cromwell, who knew even less mathematics than the prince, induced Pell to return to England, and insisted on endowing him with a handsome salary. It must be said that Pell himself was innocent of any flimflam in all this; he just kept his mouth shut and left his fortune to the ignorance and stupidity of his patrons. It is notoriously easy to make a mathematical sucker of a military man, a statesman or a politician, and that is what Pell made of the bullet-headed Oliver. One distinction led to another; Oliver sent his mathematical genius on a religious mission to Switzerland and tripled his pay. On Cromwell's death (1658), Pell, who had toadied to the royalists, consolidated his dubious position in the Restoration by taking holy orders and kissing the foot of Charles. He did not get the bishopric he coveted. Worthless or crackpot projects, such as a fantastic calendar reform, brought him further renown, if not honor. Either by dishonesty or incompetence he drifted into insolvency, and spent his last years sponging on the generosity of his tolerant and deluded friends—genius must be served. He was reputed to have done great mathematics. Had he? His friends, who couldn't tell plus from minus, insisted that he had. Pell did not contradict them; he maintained a discreet silence and a mystifying inaccessibility. This is enough of his life. There is a lot more that might suggest a diverting tale to a novelist interested in seventeenth-century reputations.

In the meantime, there is a fictional parallel to the life of Pell in the story by the Scotch satirist and journalist, Sir James M. Barrie, of the Scotch professor's devoted young wife. She had denied herself so that her genius of a husband might have undisturbed leisure to produce his masterpiece. The professor died before his wife. On searching for the masterpiece, she found it in the bottom drawer of the desk in the professor's study: a ream of virgin foolscap.

Pell's bequest to posterity was equally valueless. The legend of his mysteriously impressive mathematical reputation vanished in a mass of computational trivialities. Much of his output is carefully preserved in the British Museum. His name

245

survives in the misnomer "the Pell [or Pellian] equation." He never even saw the equation.

9. Van Schooten (?–1661)

F. van Schooten figures frequently in the correspondence of Fermat. Personally he was a decent sort. He is remembered for having taught mathematics to Huygens, Hudde,[3] Sluzze and others. He held a professorship at Leyden. His most useful work was a Latin translation (1657) of Descartes' *La Géométrie*. Profiting by Descartes' introduction of coordinates, he applied the Cartesian method to loci in three dimensions.

---

[3]For an account of Hudde's (1633–1704) work, see J. L. Coolidge, *The Mathematics of Great Amateurs* (Oxford, 1949). See also there for Van Schooten.

12

From the Correspondence
of Fermat[1]

1.　　Fermat (1636) mentioned to Mersenne questions on numbers. Says he sent to M. de Beaugrand, with the construction, to find an infinity of numbers with a certain property of aliquot parts. Letter lost. Evidently on amicable numbers.

2.　　Sainte-Croix (1636) had proposed, according to Descartes, to find two numbers which with their sum shall be sums of *exactly* 3 squares, and he gave

$$3, 3, 6; \quad 3, 11, 14; \quad 3, 21, 24;$$

saying these are all. But $11 = 4 + 4 + 1 + 1 + 1$; $14 = 4 + 4 + 4 + 1 + 1$.

　　Fermat (1636) states the impossibility of $N = x^2 + y^2 + z^2 = u^2 + v^2$ in integers or rationals.

---

[1] *Oeuvres*, 2, 4. Though the correspondence from the *Commercium Epistolicum* (*Correspondence*) of Wallis (1616–1705) could be condensed, I have given most of it in full, though sometimes paraphrased, for the occasional light it sheds on Fermat's personality. A comment after the quoted or paraphrased items will be marked by an asterisk.

3.   (Fermat to Roberval, 1636)   "Permit me to change the subject and ask you for the demonstration of this proposition which I frankly confess I have not yet been able to find, although I am assured that it is true: If a, b are rational, and if

$$a^2 + b^2 = 2(a + b)x + x^2,$$

both x and x^2 are rational.

"You cannot believe how defective the Tenth Book of Euclid is: I mean that this 'theory' has not yet made great progress and that it is nevertheless of the greatest use. I have discovered many new 'lights,' but still the least thing stops me, like the theorem I have just written to you, which seems at first to be easier to prove than it is."

*3.1   In some ways this is the most extraordinary thing Fermat ever wrote. The problem that baffled him is essentially trivial, as shown by Roberval using a simple argument by parities. This problem, proposed when Fermat was on the threshold of his great period, shows how sudden was his awakening from the "dogmatic slumbers" of his erudition inherited from the ancients. Of course he had begun to slough off some of it, as in mechanics and geometry. But here he is still just another puzzled scholar.

Fermat (1636) says he is working at what may be called the *algebra* of polygonal, pyramidal numbers.

4.   (Mersenne [1638] had asked Descartes for a rule for finding numbers having a given ratio to the sum of their aliquot parts. Fermat to Mersenne)   "As for the numbers of aliquot parts, if I have leisure, I will put in my analytic method on this subject and share it with you. I find that M. Frénicle holds himself very well hidden and will not disclose to you his

artifice. I shall not do the like, for you sufficiently know with what liberty I expose all my thoughts."

5.   (Descartes to Fermat, 1638)   "I well know that my approbation is not at all necessary to make you judge what opinion you should hold of yourself, but if it could contribute something so that you should do me the honor to write to me, I think I am obliged to confess to you frankly that I have never known anyone who made it seem to me that he knew as much as you in Geometry." (He then proceeds to disagree [wrongly] with Fermat on tangents.)

(Mersenne, 1639, *Nouvelles Pensées de Cailée*; Preface, 9–10)

"I come now to aliquot parts, which offer more difficulty of discovery than any other difficulties of mathematics; whence not many have been able to come to the end. Now the first number one has taken to work on is 120, the sum of whose aliquot parts is the double 240. As far as I know no more have been found." [2]

6.   (To Mersenne, 1640)   Magic square 17×17: take off 2 borders, a square remains; says he has no general rule for such artificial squares. Answering Frénicle. Then "solid" squares. Much on magic squares.

7.   (To Mersenne, 1640)   Perfect numbers; several short cuts to find such. "There are none of 20 or 21 digits, destroying the conjecture of those who have believed there is one in every '*dixaine*,' e.g., 1 in 1 to 10; in 10 to 100; 100 to 1000 and so on.

---

[2] The general problem appears to be the discovery of a number N the sum of whose aliquot parts is equal to a stated multiple or a stated submultiple of N. Several special cases were found, but nothing generally usable.

From after 10,000,000,000,000,000,000, to the next power of 10 there is none.

"To see whether M. Frénicle uses tables, propose these [repeated from letter of 1636]

> *Find a right triangle whose area is a square.*
> *Find 2 biquadrates whose sum is a biquadrate.*
> *Find 4 squares in arithmetical progression.*
> *Find 2 cubes whose sum is a cube.*

"If he replies that up to a certain number of digits he has proved that these problems have no solution, be assured that he proceeds by tables." (All are impossible in integers.)

8.   (Fermat to Mersenne, 1640)   Magic squares again. Fermat says to tell Bachet that his doubts are groundless; for ten years he has had a general method for finding all even squares to infinity, and had given examples of squares of order higher than those of Bachet. He says M. Despagnet can testify. He says he has not thought of finding out the number of squares for a given order; and "avows" that his methods are not exhaustive. Has one of order 22, but as the courier is leaving, he hasn't time to write it out. Also something about magic cubes. More on perfect numbers, all deducible from Fermat's "lesser theorem."—Calls them the foundations of the discovering of perfect numbers.

His rules: "the numbers less by 1 which proceed by 'double progression' [2^n, $n = 1, 2, 3, \ldots$]:

1	2	3	4	5	6	7	8	9	10	11	12	13
1	3	7	15	31	63	127	255	511	1023	2047	4095	8191

Call second line the roots of perfect numbers, because, whenever they are primes, they produce perfect numbers. Write above these in natural order 1, 2, 3, 4, 5, etc., which are called their exponents.

"That supposed, I say that:

1. "Whenever the exponent of a radical number is composite, its radical also is composite. For example, 6 the exponent of 63 is composite; I say that 63 is also composite.

2. "Whenever the exponent is a prime number, I say that its radical less 1 is 'measured by' the double of the exponent. E.g., since 7, the exponent of 127, is prime, I say that 126 is divisible by 14.

3. "Whenever the exponent is a prime number, I say that its radical can not be measured by any prime number except those which exceed by 1 a multiple of the double of the exponent or the double of the exponent. E.g., since 11, the exponent of 2047, is prime, I say that it can not be measured except by a number exceeding 22 by 1, that is, 23, or by a multiple of 22, plus 1: indeed, 2047 is divisible only by 23 and by 89.

"These are 3 very beautiful propositions which I have discovered and proved not without difficulty: I may call them the foundations of the theory of perfect numbers.

"*Interesting note:* You or I have equivocated [this part of letter lost] by certain characters of the number which I thought perfect, as you will easily see since I took 137438953471 for its radical, which I have since nevertheless found, by the abbreviation drawn from my third proposition, is divisible by 223; as I knew at the second division I made, for the exponent of this radical being 37, whose double is 74, I began my divisions by

149, greater by 1 than the double of 74; thus, continuing, by 223, greater by 1 than the triple of 74; I found that the radical in question is a multiple of 223." (He asks Mersenne to get Frénicle to assist him in the numerical computations.) "In any event, I should like you to get Roberval to join his work to mine, since I find myself pressed by many occupations which leave me only very little time to 'attend' to those things."

9.   (Roberval to Fermat [last part of letter], 1640)   "I almost forgot to speak to you about numbers, of which you have already discovered some admirable properties, containing great mysteries; but to discover them better, would demand several [collaborators], in agreement and without jealousy, and whose genius was naturally suited to this speculation, which is very difficult to undertake. If this subject pleases you, or one of those of whom I have spoken above, I also would take pleasure to consider it more particularly, hoping that you will share your 'inventions' with me, so that I may [collaborate]."

(Fermat's reply)   "After having thanked you for your civilities and having protested that I should be ravished to have occasions to please you ..." he passes on to geometry (tangents); then to integer or rational sums of two squares. No $x^2 + y^2$ (x, y) coprime is divisible by a prime $4n - 1$.

10.   (To Frénicle, 1640. The first statement of $2^{2^n} + 1$ being prime for all n is wrong.)

After several minor theorems: "But here is what I admire the most: it is that I am almost persuaded that all the numbers of the sequence, increased by 1 whose exponents are numbers of the 'double progression,' are prime numbers, as

$$3, \quad 5, \quad 17, \quad 65537, \quad 4294967297$$

and the following of 21 digits

$$184 \quad 467 \quad 440 \quad 073 \quad 709 \quad 551 \quad 617$$

etc.

"I have not an exact demonstration, but I have excluded so great a quantity of divisors by infallible demonstrations, and I have such great insights, which establish my thought, that I would have difficulty to contradict myself.

"For altho' I reduce the exclusion to most numbers and although I have probable reasons for the rest, I have not yet been able to demonstrate necessarily the truth of my proposition, of which nevertheless I do not doubt at all at this hour what I already did. If you have an 'assured' proof, you will oblige me by communicating it to me: for after that, nothing will stop me in these matters."

*10.1   Fermat stated that he *thought* the proposition true, but *never anywhere claimed that he had proved it*. It is time that the erroneous statements on this matter were corrected in some accepted histories of mathematics—even at the cost of printing *the whole* of what Fermat said in his own language.

11.   (To Frénicle, 1640)   States in effect his "lesser" theorem: If the integer a is not divisible by the prime p, then $a^{p-1} - 1$ is divisible by p.

12.   (To Mersenne, 1640)   "If I could once grasp the fundamental reason why $3, 5, 17$, etc., are prime numbers it seems to me that I could find very beautiful things in this matter, for already I have found some marvelous things which I will share with you, after I have seen your reply and that of M. Frénicle."

*12.1   "If I could once grasp," etc. This is one of the profoundest things Fermat ever said.

*12.2   "Every prime $4n + 1$ is once only a sum of two squares, and once only the hypotenuse of a right triangle. Its powers are sums of 2 squares." (Gives numbers of ways this can happen for the powers. Finds how many solutions of $n = x^2 + y^2$, n given.[3])

13.   (1641) Mersenne asks for return of a letter to Fermat as he had not kept a copy.

14.   Given 2 integers, find 2 right triangles which have the given ratio of the 2 integers. There are 4 rules for this.

Triangles whose least sides differ by 1: 3, 4, 5;   20, 21, 29.—General rule.

Other triangles in which least side always differs by 1 from each of the others by a square, e.g., 20, 21, 29.

To find a number which shall be as many times as one wishes (a prescribed) number of times, and no more, a polygonal number. (Proposed 1641 to Frénicle.)

Solution of his area of triangles problem: Let a, b be the numbers expressing the given ratio. The triangles: $a^2 + b, a - b$; $b^2, a - b$.

---

[3] Fermat's general method, that of infinite descent, is profoundly ingenious, but has the disadvantage that it is often extremely difficult to apply. In the particular theorem for which Fermat invented the method, it is required to prove that every positive prime of form $4n + 1$ is a sum of two integer squares. From the assumption that the theorem is false for some such prime p, Fermat deduced that it is also false for some smaller prime of the same kind. Descending thus he proved on the assumption of falsity that 5 is not a sum of two squares. But $5 = 1^2 + 2^2$; hence the theorem.

Euclid used a similar type of reasoning in his proof of the infinity of primes: a square of positive integers cannot decrease indefinitely.

15.   Many other examples of right triangles with special relations between the sides, for which he says he has rules "to infinity."

16.   (Frénicle to Fermat, 1641)   "I know that the algebra of this country [France] is not adapted to resolve these questions [of Diophantine analysis], or at least nobody here has found the method of applying it. For this reason I believe that you have recently 'devised' some particular [kind] of analysis to 'search' in the profoundest secrets of numbers, where you have found some ability to serve in this effort from which you have been accustomed to employ in other usages. ... The methods [for certain triangles] you give are very beautiful, and you have a method to dispose your rules, such that it gives them a certain grace." (Always complimentary, admitting that he is not in Fermat's class, but "by the grace of God" has found some things for himself, especially in polygonal numbers.)

17.   (Fermat to Mersenne, 1642)   "Remember to communicate the writings of M. Frénicle, for the love with which I have worked at numbers, and I am assured that I will someday persuade you that my work has not been useless."

18.   (Fermat to Mersenne, 1642)   "Don't know what status I shall have in the mind of M. de la Chambre since the commission of Castres has succeeded so badly."

19.   (To Carcavi, 1643)   "Lest you should accuse me of not sending you anything of my inventions, I send you three numbers among several more whose aliquot parts 'make the multiple.'

"The following number is *sous-triple* $[\frac{1}{3}]$ of its aliquot parts

14   942   123   276   641   920.

The next is *sous-quadruple* [$\frac{1}{4}$]

$$1 \quad 802 \quad 582 \quad 780 \quad 370 \quad 364 \quad 661 \quad 760,$$

and also this one

$$87 \quad 934 \quad 476 \quad 737 \quad 668 \quad 055 \quad 040.$$

Then on this matter here are two which I have chosen among my *sous-quintuples* [$\frac{1}{5}$], the first is the product of the following numbers

8 388 605, 2801, 2401, 2197, 2187, 1331, 467, 307
289, 241, 125, 61, 41, 31,

and the other is the product of

134 217 728, 243, 169, 127, 125, 113, 61, 43, 31, 29, 19, 11, 7.

Here is a *sous-double* [$\frac{1}{2}$] of its parts, that I have discovered, which multiplied by 3 gives a *sous-triple* [$\frac{1}{3}$]:

$$31 \quad 001 \quad 180 \quad 160.$$

I have found the general method for finding all possible [I have a quantity of others], which I am assured will astonish M. de Roberval and even the good Father Mersenne also; for there is certainly nothing in the whole of mathematics more difficult than this, and except M. de Frénicle and perhaps M. Descartes, I doubt whether anyone knows the secret, which however will not be for you any more than a thousand other inventions [on] which I may engage you on another occasion."

20.　　(To Mersenne, 1643)　　Mersenne had asked Fermat for a method of finding whether a number is prime or composite and doing it in a day. Fermat replied by an example:

$$100\ 895\ 598\ 169 = 898\ 423 \qquad 112\ 303,$$

both factors primes.

256

More on factorization, but no *general* methods. More on triangles.

21.   Asserts that he always has solutions for the problems he proposes. They have given "a bad impression of me as having proposed an amusement and a useless work. ... I'll send him whatever solutions they ask."

22.   (To Pascal, 1654, on $2^{2^n} + 1$)   "This is a property on whose truth I answer you. The proof of it is very difficult, and *I tell you straight out that I have not yet been able to find it fully; I would not propose it to you to seek if I had come to the end.*

"This proposition serves for the invention of numbers which have a given ratio to their aliquot parts, on which I have made considerable discoveries. We shall speak of them another time."

23.   (Fermat to Pascal, 1654)   "I hope you will send Saint-Martin an abstract of all I have invented, which is 'quite-considerable,' on numbers. You will permit me to be concise and to make myself understood only to a man who understands all at half a word.

"What you will find the most important concerns the proposition that every number is composed of one, of two, or of three triangles; of one, of two, of three or of four squares;— pentagons, to 5, hexagons 1 to 6; and so to infinity. [Communicated to Mersenne in 1636; see Chapter 3 here.] A prime $4n + 1 = x^2 = y^2$; $5, 13, 17, 29, 37, \ldots$.
Given such a prime, say 53, to find its x^2, y^2.
A prime $3n + 1 = x^2 + 3y^2$: $7, 13, 19, 31, 37, \cdots$.
Primes $8n + 1$, $8n + 3$, $= x^2 + y^2$; $11, 17, 19, 41, 43$.
No right triangle in numbers whose area is a square.

"That will be followed by the invention of several propositions which Bachet didn't know and which are lacking in Diophantus.

257

"I am convinced that when you shall have seen my fashion of proving these propositions, it will seem beautiful to you, and will give you the opportunity to make many new discoveries.

"If time remains to me, we shall speak next of magic numbers, I shall recall my old species on this subject."

24.   (Pascal to Fermat, 1654)   Writing from Paris, Pascal replied that Fermat's arithmetical theorems were far beyond him: "I am capable only of admiring them, and I beg you very humbly to occupy your first leisure in proving them. All our gentlemen saw them last Saturday and admired them with all their heart; one cannot easily bear waiting for such beautiful and such desirable things."

25.   (Fermat's reply to Digby, 1657)   "I dare to say to you, with respect and without lowering the high opinion which I have of your nation, that the 2 letters of my Lord Brouncker, although obscure in my opinion and badly translated, do not at all contain a solution." He is "not trying to revive the jousts and old blows of lances the English formerly carried on against the French"... but he reminds Digby that chance and good luck sometimes arise in the combats of science as in others.

"I would moreover be delighted to be undeceived by this ingenious and learned lord, and, to assure him that our combat will not at all be to a mortal one, in the following question, that I am going to propose to him, I relax the rigor of my previous questions which asked for only integer numbers: it will satisfy me if they are rational in the fashion of Diophantus.

"Here is the new question, either for my Lord Brouncker or for Mr. Wallis, which I write in Latin according to your order:

"*Given a number composed of two cubes to divide it into two other cube numbers.*"

(E.g., $28 = 1 + 27$. Divide 28 into other cubes [rational], and give the general solution of the proposition 11.)

(Mentions unique solutions of $x^2 + 2 = y^3$, $x^2 = 25$.) "Frénicle could scarcely believe this, and found it too bold and too general. But, to increase his astonishment, I say that if one seeks a square, which added to 4, makes a cube $[x^2 + 4 = y^3]$, he will find only $x^2 = 4$, $x^2 = 121$. For 4 added to 4 makes 8, a cube, and 121 added to 4 makes 125, which is also a cube. But, after that, the whole infinity of numbers cannot furnish a third with the same property.

"I don't know what the English will think of these negative propositions and if they will find them too bold: I await their solution and that of Mr. Frénicle, who has not replied to a long letter Mr. Borel delivered to him from me. By which I was surprised, for I replied exactly to all his doubts, and I made him some question of my "*chef*" of which I await the solution."

26.   (Fermat to Digby, 1658)   "After having received Mr. Wallis' letter, I am always surprised how he constantly is contemptuous of what he does not understand. Questions on whole numbers are not to his taste. He imagines that I know nothing of the centers of gravity of infinite hyperbolas, etc.

"I answer him succinctly: First he says I make a great to-do of negative propositions, such as $x^2 + 2 = y^3$ ($x^2 = 25$) and $x^2 + 4 = y^3$ ($x^2 = 4, 121$) and pick no more.

"I reply that I make no '*cas*' of *all* sorts of negative propositions; those which he [Wallis] reports and an infinity of such a nature are only amusements for a three-day arithmetician, and their reason is immediately known *etiam lippis et tonsoribus* [Horace, Sat. I 6, 3: even to the purblind and barbers]. Thence he infers that little account should be taken of all sorts of negative propositions, see, Sir, what logic! But I wish no other proof that those which I have proposed to you are of a high quality and worthy of being investigated, it is because neither he, who so greatly esteems himself, nor even

259

Mr. Frénicle, whom I place above him, without being prejudicial to him; and this last who knows marvelously the most hidden secrets of numbers, has not misapprehended them.

"But because integer numbers do not please Mr. Wallis, here is another which he may occupy himself with and in which I do not exclude fractions

"There is no right triangle whose area is a square.

"And, to make him see the lack of knowledge of this sort of questions sometimes makes him conceive a greater opinion of his powers than he might reasonably have of them, he has no doubt that my Lord Brouncker can solve the two following questions:

"To resolve a given cube into 2 rational cubes;

"Give a number which is composed of the sum of two cubes, to divide it into two other rational cubes;

"I reply to him that he may, by chance, not deceive himself in the second, although it is difficult enough, but that for the first it is one of my negative propositions which neither he nor Lord Brouncker perhaps can demonstrate so easily. For I maintain that *there is no cube in numbers which can be divided into two rational cubes.* [For more on Wallis, see Chapter 11 here.]

"For the second question, it is not of extreme difficulty, and to testify that I wish even to propose it to him in the easiest case of taking a small number, I will be satisfied if he or my Lord Brouncker divide the number 9, which is composed of the 2 cubes 8 and 1, into two other rational cubes. If he rejects this proposition, which is not among the most difficult, I would not dare to propose it to them neither in integers nor in fractions."

For a rule Wallis proposed on squares, "I don't know why he should doubt that this invention should seem difficult to me, since any novice in algebra would find the rule immediately.

But my question *in integers* is so far above those trivial rules that Mr. Frénicle has judged it worthy of occupying himself with it, and that is to say everything." [Then follows a geometrical question.]

27.　　On several occasions Fermat proposed "challenge problems" to the mathematicians of Europe and elsewhere if there were any. These stirred up a great deal of interest and some animosity when they proved too hard for the challenged. A first type, "multiply perfect numbers," was suggested by the current interest in perfect numbers and provided one kind of "generalization" of such numbers. It is not completely solved even today, though numerous instances of the numbers $\sigma(x)$, $\Sigma(x)$ have been found since the seventeenth century, especially in our own times. The mathematical importance, if any, of σ, Σ today may be slight. But the questions it raises are difficult, so much so that this particular challenge has been dismissed by those it baffles as trivial. Here it is:

Let $\sigma(x)$ denote the sum of the divisors of the positive integer, x, *x included*, and Σx the like *x excluded*.
As specimens of the challenges:

(A) *First problem (1647)*: To find a cube, which, with its divisors, gives a square. Example, Σx: The divisors of $7^3 = 343$ are $1, 7, 49$. With 343 these give $400 = 20^2$. Required another solution.

Sent to Wallis, Frénicle, Schooten. Frénicle gave four solutions, e.g.,

$$x = 2 \cdot 3 \cdot 13 \cdot 41 \cdot 47, \qquad y = 2^7 \cdot 3^2 \cdot 5^2 \cdot 7 \cdot 13 \cdot 17 \cdot 29.$$

He said greater could be found but the calculations were too

tedious. He noted $\Sigma 7^3 = 20^2$, $\Sigma(20)^2 = 31^2$.

261

(B) *Second problem*: To find a square which, with its divisors, gives a cube y^3.

For (B), viz. $\Sigma(x^2) = y^3$, Frénicle gave

$$x = 7 \cdot 11 \cdot 29 \cdot 163 \cdot 191 \cdot 439, \qquad y = 3 \cdot 7 \cdot 13 \cdot 19 \cdot 31 \cdot 67.$$

Four further problems were proposed to the mathematicians of Europe:

(P) $\Sigma(x) = 5x$, so that $\sigma(5) = 25x$. Frénicle gave the first three solutions with $44, 45, 55$ digits respectively. Further solutions were sought.

(Q) $\Sigma(x) = 7x$ so that $\sigma(7x) = 49x$.

(R) To find a "central hexagonal number" [I need not state the definition] which is a cube.

(S) To find $2, 3, 4$ or more consecutive central hexagonal numbers whose sum is a cube. Those things, though not easy, are of no interest today. Frénicle disposed of (P), (Q), (R).

(A), (B), (P), (Q) were suggested by the contemporary interest in perfect numbers. Such questions gave rise to the search for "multiply perfect" numbers. The first example goes back to R. Recorde who, in his *Whetstone of Witte*, 1657, noted that $\sigma(120) = 2 \times 120$. Mersenne (1631) proposed to Descartes to find another like this. Descartes could not, and disposed of the challenge by saying that this kind of problem is a typical contest of grubbing endurance. If that is really how he felt about the challenge he should have ignored it entirely. Mersenne (1634) gave 30240 as a solution of $\sigma(x) = 3x$, and proposed $\sigma(x) = nx$. Fermat's reply is lost, but he had sent a correspondent a rule for finding an infinity of solutions of the problem. For $\sigma(x) = 2x$ gave the fourth solution, Frénicle the fifth. Descartes discussed $\sigma(x) = 3x$, $\sigma(x) = 4x$. In a letter to Carcavi, 1643, Fermat gave further solutions, and two solutions

"I have next considered certain questions which, although negative, are not too difficult, altho' the method of descent is quite different from the preceding, as it would be easy to test. Such are the following:

$$x^3 + y^3 = z^3 \quad (\text{impossible})$$

$$x^2 + 2 = y^3 \quad (\text{to prove there is only one solution})$$

$$x^2 + 4 = y^3 \quad (\text{to prove there is only one solution})$$

$$2^{2^n} + 1 \qquad \text{all primes (false, proof not claimed)}$$

This last question is of a very subtle and very ingenious research and, although it is negative, since to say that it is a prime number, is to say that it cannot be divided by any other number.

"I put in this place the following question of which I have sent the demonstration to M. Frénicle, after he acknowledged to me and has even testified in print that he was unable to prove it:

"There are only 2 numbers, 1, 7 which being less by 1 than twice a square make a square of the same kind, that in which are twice a square minus 1: $2x^2 - 1 = (2y^2 - 1)^2$.

$$x = 1, y = 1; \; x = 5, y = 2.$$

"Bachet glorifies himself, in his commentaries on Diophantus, to have found a rule in two particular cases. I give the general rule for all sorts of cases and determine whether a double equation is possible or not.

"I have then re-established most of the defective propositions of Diophantus and have done those which Bachet avows he did not know and most of those in which it seems that Diophantus himself has hesitated, of which I shall give proofs and examples at my first leisure.

"I admit that my invention for discovering whether or not a given number is prime or not is not perfect, but I have many ways and methods for reducing the number of divisions and by these much diminishing and abridging the usual labor. If Mr. Frénicle [tells me] that he has mediated on this, I should esteem what would be a considerable help for savants.

"The question which has occupied me without my yet finding any solution is the following, which is the last in the Book of Diophantus, *De Multangulis Numeris* [*On polygonal numbers*].

"Given a number, to find in how many ways it may be "multangulus" [polygonal]. The text of Diophantus being corrupt, we cannot guess his method; that of Bachet does not please me and it is too difficult for large numbers. I have indeed found a better, but it still does not satisfy me." (Still unsolved; see Chapter 3 here.)

In the course of this proposition must be sought the solution of the following problem:

"To find a number which may be a polygon as many times as one wishes, and to find the smallest of those which satisfy the question.

"There, summarily, is the count of my reveries on the subject of numbers. I have written it only because I anticipate that leisure to put out at length all these demonstrations and these methods will be lacking to me; in any event this indication will serve savants to find for themselves what I do not extend, principally if Messers. de Carcavi and Frénicle participate in some demonstrations *by descent* which I have sent them on the subject of certain negative propositions. And perhaps posterity will take it good for [having been made aware] that the Ancients did not know everything, and their telling may pass into the mind of those who come after me '*pour traditio lampadis ad filios*' [to hand on the torch to the sons], as spoke the great Chancellor of England [Bacon] following whose sentiment and to whose legend I add 'Multiper-transibunt et augebitur scientia' [Many shall run to and fro, and knowledge shall be increased. Daniel, 12, 4]."

29.   (Pascal to Fermat, August 1659)   "When I knew that we are closer to one another than formerly, I couldn't resist a friendly plan on which I have begged M. de Carcavi to be the mediator: in a word I aim to meet you and converse some days with you; but, because my health is hardly stronger than yours, I dare to hope that this consideration will make you grant me the favor of half the road [halfway], and that you will oblige me by indicating a place between Clermont and Toulouse, where I will not fail to be toward the end of September or the beginning of October.

"If you do not take this means, you run the chance of seeing me at your house and having two sick people there at the same time. I wait your news with impatience. ..."

30.   (Pascal to Fermat, August 10, 1660)   Begs off on account on of ill health, otherwise would hasten to accept, even to "flying" to Toulouse, "and I wouldn't have suffered that such a man as you should take a step for such a man as I. ... To speak frankly to you about mathematics, I find it the highest exercise of the mind: but at the same time I know it to be so useless that I make little difference between a man who is only a mathematician and a skilled artisan. Also I call it the most beautiful trade in the world, but then it is only a trade and I have often said that it is good *pur faire l'essai*, but not *l'emploi de notre force*. ..." Goes on about his health: "I can't walk without a cane, nor can I ride a horse, even I cannot make more than three or four leagues *en carrosse*. It was thus that I came to Paris 22 days ago. The doctors have prescribed the waters of Bourbon for the month of Sept.," etc.

269

13

An Age to Remember

Fermat's office as a king's councilor gave him many opportunities in his journeys from court to court to observe the degradation of the peasants, the decadence of the nobility, the evidences of political and administrative corruption everywhere, the callous, deliberate brutality of the military, and the total indifference of the socially elite to the bestial existence of their inferiors. Fermat was not an inhumane man, but as a dutiful and loyal subject of his God and his king there was little he could do. When he was inclined to intercede for the poor and the oppressed he had to move with extreme caution, or he would soon have found himself dispossessed of his office and his one chance to alleviate the inhuman suffering. From a jurist, open expressions of sympathy were improper and dangerous; only a few trusted friends received Fermat's confidences. Nevertheless he managed to publicize certain abuses openly. Having done so he resumed his unquestioning loyalty to his king, as was his duty. He will speak for himself (in the next chapter) after we have noted some of the conditions under which the French intellectuals of the seventeenth century lived and created memorable work.

270

If we can detach ourselves in time and place for a few moments and compare the culture of seventeenth-century France with that of our twentieth century almost anywhere in the civilized world we may note certain similarities in the two. Both exhibit an explosive change in science, particularly in the physical sciences and mathematics. In each culture there is an all but continuous succession of costly wars and their accompanying burden of ruinous taxation to pay for them, or to mortgage the future to promise an illusory solvency. There is one important distinction however: our debts, we sincerely hope, will be liquidated; the French ignored and forgot theirs. One thing is lacking in the parallel: an alliance of dehumanized tyrants and able despots to prepare, in spite of themselves, a catastrophic social upheaval that would sweep them and their followers to a common ruin.

Unlike our own time, France was dominated by three masterly architects of revolution concerned almost wholly with their personal aggrandizement:

> Cardinal Richelieu 1585–1642
> Cardinal Mazarin 1602–1683
> J. B. Colbert 1619–1683

(Our own aspiring tyrants of recent times, Hitler, Mussolini, Stalin, of a different stripe, were eliminated before they could accomplish their purposes. Though they wrought great damage they died frustrated, two suffering violent deaths. Not so the three Frenchmen; they succeeded in ruining their country. All died peacefully in bed.)

Richelieu, the notorious cardinal and ruthless tyrant, is familiar to every adolescent devotee of historical romances. Steadfastly pious and invincibly ignorant, he was an ornament to his high office.

271

Mazarin, less well known, was a slick importation from Italy. If he was married to the widowed queen—as maybe he was—it was never publicly acknowledged. He is tenderly remembered by sentimental ladies. Professionally he was one of the greatest virtuosos of the intricate art of taxation that has ever looted and bankrupted a nation. Even some of our own artists might learn from Mazarin. Nor was he backward in double-dealing and unscrupulous diplomacy; he bequeathed France a crazy patchwork of unstable treaties and ephemeral alliances. But in spite of, or perhaps because of, all the hatred his tax exactions had aroused against him, he died an almost indecently rich man. Unlike Richelieu and Colbert, Mazarin had some of the softer human qualities. He remained faithful, in his fashion, to the queen who had pulled him up, and is said to have loved her for herself and not merely for what she brought him. To console himself for the lack of a faithful and devoted consort, he adopted seven beautiful, talented and nubile young girls. These, by the accepted euphemism of the time, were the Cardinal's nieces. He transplanted the lot to Paris, where he married them off to assorted counts and dowered them all handsomely. The good cardinal, strangely enough, had no nephews.

Of the three architects of revolution, Colbert was by far the ablest. He is remembered with respect by intellectuals for his promotion and fostering of the French Academy of Sciences. If he had any fluid in his veins it must have been ice water. Madame de Sévigné, whom we shall meet later, feared him. She called him "the North." If he possessed a human side or a likable weakness he managed to conceal it. But he did have one overmastering passion, France. If all went as he planned and plotted he would make his France the mistress and leader of Europe. He had been indoctrinated in the French War Office. Seeking a more congenial field for his talents, at the age

mostly for rabbits, in the fields of the peasants, trampling down the crops. On a beggarly scale their ridiculous gambols were a parody of the aristocratic English fox hunts. Any game the peasants themselves might trap, though an essential part of their sustenance, was surrendered to the gentlemen.

Louis was not too well pleased with this scum of "decayed nobles of the sword." Their vulgar wrangling for recognition and preferment disgusted even him. He fed and clothed them to flatter his vanity and amuse the ladies. This role of mountebank was galling enough to the pride of the once admired and envied nobles of the sword. To rub in their humiliation, because they were of noble birth they were kept in a proper state of ignorance, being permitted to spend less time than a commoner at the university. One sop to their vanity was retained: if one not of the nobility challenged a gentleman to a duel he was hanged with the customary refinements, win or lose. An unpopular gentleman, on the other hand, did not need any challenge to be eliminated; he was easily liquidated in the unlighted streets.

The nobles waited their opportunity to recoup their losses of revenue and respect. They got their chance when civil war (the "Frondes," 1648–1653) erupted. Anarchy, or a fair imitation of it, broke out in Paris and the provinces. The nobles of central France recaptured their feudal privileges of the sword by brigandage. They set up their own courts to try recalcitrant peasants, examining thousands, condemning many, but in their congenital incompetence letting most escape. For "political" reasons Louis pardoned the nobles responsible for the fiasco. All was once more serene—for the moment.

A fair criterion for the degree of material civilization of a country is the condition of its roads and the available means of transportation. Judged by this, seventeenth-century France had

277

barely emerged from the Middle Ages. Travel by river barges where obtainable was superior to that by land except in the few instances where short stretches of the old Roman roads were not yet obliterated. Some of these were still in fair condition and usable, but they connected only the larger towns. To branch off and seek the smaller towns and villages was an adventure not to be lightly undertaken. In wet seasons the trails, "pig tracks," from place to place ranged almost at random over fields, moors and swamps. Nothing resembling a purposeful connecting link existed, and such routes as the bewildered travelers followed were deeply rutted successions of chuck holes and mud bogs. Another nuisance impeded travel: the better roads were infested by highwaymen or self-appointed toll keepers. These predatory gentry were recruited from the ranks of the practically destitute nobility. Denied an opportunity to work on account of their rank, they lived by brigandage, thieving and pillage. They had no legal right to impose tolls, yet clamped them down anyway for all the wretched traffic could bear.

Travelers who had the price hired armed bodyguards as outriders who fought small pitched battles to see their charges safely through. The brigands holed up in their crumbling castles, whence they issued at the slightest hint of safe loot. Richelieu cleaned up this particular nuisance by ordering the demolition of many castles. Any haphazard repairs of the broken-down roads were carried out by forced labor for which the workmen—peasants—received not a sou of compensation. Nor any rebate on their taxes. They also supplied all necessary materials gratis. As in everything else, whatever law was supposed to exist was exclusively for the privileged few.

The upper classes, including the officials and dignitaries of state and Church, of course traveled in such security as was obtainable and in comparative comfort. Two specimen journeys of the nobility will suffice: that of La Grande Mademoiselle (1627–1693), daughter of Gaston, Duke of Orléans; and that of Richelieu. What amounted to a small private army as bodyguards conducted the lady safely along through a fringe of uninvited barbarians dogging her coach. Her armed escort also protected the convoy of carts and pack mules following the coach; there was no attempt to rob them. She made the journey by short stages, stopping after dark at castles along the way and replenishing her strength with stupefying gorges of eight-course dinners of twelve dishes to the course—quite a contrast to what her escort and the country people got. To avoid the worst of the roads she took a barge down the Loire. All in all a not unpleasant outing. But the evenings were dull; bedtime was at nine; at dawn they were on their way again.

Even when escorted by guards, travel in many sections was extremely hazardous. There was no guessing when one of the roving bands of armed men might take a long chance and attack in broad daylight. Mademoiselle proceeded with caution. But she kept going. Accommodations along the way were sometimes in tumble-down houses and shacks barely fit for cattle. It was worst at night. Huddled together to keep warm in the cold and wet dark, the travelers, nearly exhausted as they were, dozed and longed for the cheerless daybreak.

It was a season of swollen rivers and floods making the roads marshes and swamps. The only means of getting anywhere was to cut through the dangerous woods and over the moors by random tracks, all filthy. Mademoiselle, being very thirsty one noon, sampled the water. It apparently was sewage; she went on thirsty. Aiming for Paris, which she wished to by-pass, she lost her way and blundered into a full-scale insurrection of fighting peasants. At Sedan the river was in flood, so she abandoned the coach, mounted her horse and headed for Lyons. Here we recognize a name that was to become famous in the senseless butcheries of World War I, Vimy. There the Archbishop of Lyons welcomed her to his house, entertained her, gave her his blessing and waved her on her perilous way.

It must be admitted that Mademoiselle, though a scion of the hated and decadent aristocracy, had guts. She continued her journey, undeterred by washed-out bridges, floods, incipient insurrections and discomfort in general. France could have used more like her, both women and men.

Our second distinguished traveler is Cardinal Richelieu. Compared to Mademoiselle he had it soft. He was transported by two dozen stalwarts in a huge wooden litter upholstered in crimson and gold, his favorite colors. He was on his way back to Paris after inspecting his deviltries in the south. He was ill and in great pain. To ease any jolts, the litter was preceded by a corps of men carrying planks, for building a ramp to the entrance of any dwelling where the cardinal deigned to put up. It would have been a delightful outing had not his Eminence been in constant and severe discomfort from a rotting shoulder. Neither salves nor prayers alleviated his suffering. (The sanitary offices of the lowly maggots had not yet been recognized and these busy little surgeons were stupidly scooped out and discarded by the wayside.) The cardinal's naturally irascible temper became simply vile, as the country folk and litter bearers learned to their nauseated and disgusted distress.

He was in no immediate danger of robbery or murder. His large escort of professional soldiers, tough and well paid for their care, handed him as tenderly as a rotten egg and discouraged any loiterers from attempting to share the cardinal's bounty. With a little courage and a dash of daring the escort might have cured France then and there of its malignant cancer by a bold surgical operation. But the bought soldiers preferred to play nursemaid to the grievously sick man, thereby ensuring their wages and leaving surgery to the maggots and spiritless brigands with little stomach for the sight of blood. So his Eminence reached Paris intact, in a murderous mood and ripe for raising hell unlimited. He was a tough old devil.

We must now return to the Sun King.

As a tribute to Louis XIV's administrative genius, so often extolled by his eulogists and admirers, not one year of his reign passed without a revolt of the lower classes in the towns or in the country or in both. Most of the discontent was due to crushing taxation needed to finance four years of vicious civil war grafted onto thirty years of ruinous and futile foreign wars. The unchecked extravagance of the king and his court also helped to bleed the country white. When foreigners, as well as Frenchmen, hesitantly tried to point out what taxation was doing to the country, the king, to justify an arbitrarily imposed "punitive" tax, replied that he had "the right and the power to do it."

His remedy for unpopular taxes was more of the same. A direct consequence of the ever-increasing burden was the rapid deterioration of agriculture and hence of the standard of living. The land, allowed to go uncultivated, became barren. Why should the peasants break their backs for the right to starve? The tax sharks were voracious enough, but the protected landlords surpassed them in rapacity. As much as three-fourths of whatever a peasant produced was exacted as rent by his overlord; the remaining fourth could not be stretched to feed a man and his family. Creeping starvation gradually degenerated the human stock till men could no longer do a fraction of a normal day's work. There was one gratifying outcome of the peasantry's decline: "the buzzards"—the landlords used up their reserves of fat and began to feel the sharp nip of hunger.

281

They whined that after supplying "the king's tables," they had almost nothing for themselves. They kept going by seizing or stealing whatever little the peasants had left. To keep the royal larder well stocked the king applied his panacea of more taxes. These the peasants raised by parting with half their already severely reduced ration to the buyers and sellers of food, swindlers all, who pocketed most of what they collected. Morally, socially and economically France was a cesspool of corruption. Yet it survived and became even more corrupt. Humanity is unbeatable.

The year 1662 is memorable in the annals of infamy. Louis had justified his disciplinary punitive tax by his divine right and power to tax as he pleased. The Boulonnais refused to pay the tax. The peasants had taken all they could stomach of the divine right of Louis to starve them to death. A diet of nettles and roots, eked out by the bark of trees and clay, might keep men moving but it had evidently been insufficient for the women and children, whose dead bodies cluttered the road-sides. King or no king, the sight of these corpses was too much for the peasants. They revolted, to the tune of 6000, of whom 3000 were taken prisoners by the king's loyal and brutal troops. What was to be done with the captives? Colbert, with his usual sagacity, saw the solution at a glance. It would not do to permit the rebels to be judged by the rural courts; the soft-hearted judges might let some off with a mild sentence or none. To make sure that justice was done, Colbert prepared sentences for a majority *before* they were tried. Then he ordered his loyal troops to dispose of them by hanging or breaking on the wheel, "to encourage the others." The galleys were short of slaves at the moment; Colbert gave some 400 life sentences to the galleys. For his prompt and efficient handling of the revolt,

Colbert received a perfunctory nod from Louis and the heart-felt curses of the common people.

Thus Colbert. What of Richelieu in a similar situation? As might be expected of such a master, he acted quickly and effectively. A revolt of the peasants, backed by the common people, had erupted (1639) in Normandy. They refused to pay any taxes imposed since Henry IV. It became an open season for tax collectors, one of whom was murdered in Rouen. The gabelle (salt tax) was particularly resented. The tax collector's house was stripped. Richelieu reacted by sending an army of 4000 trained troops to put down the rebellion. That settled it. All rebels captured were broken on the wheel. To rebuke Rouen for laxity in repressing the revolt, Richelieu dissolved its Parliament and deprived the town of its lawful privileges.

The rebels recognized "the foreigner" (Mazarin) as the author of their misery and threatened his life. The threats evaporated in curses which the cardinal disregarded except for increasing his bodyguard.

Later a more impressive rebellion broke out in Brittany, where 25,000 peasants attempted to set up a "peasants' code," a sort of proto-democracy. It was quickly knocked over by Mazarin, who imported 10,000 barely civilized savages from the army of the Rhine. These brutalized barbarians proceeded to kill and plunder at will. Undisciplined, ill paid, if at all, they lived off the land, lifting what they wanted wherever they found it. A plague of locusts could not have done a cleaner job. Nothing edible was overlooked, as testified in a firsthand observation by Madame de Sévigné (1626–1696), noted as a letter writer, and notorious for her somewhat morbid affection for her son and daughter, particularly the latter. On her gadding about the provinces she observed the camps of the dehumanized soldiers and blithely reported their diversions:

283

"The other day they put a little child on the spit." She then continues with chatter about her beloved and pampered brats. On another occasion she reported seeing four spitted children laid out waiting their turns on the grill. When "the day of glory" arrived for the lower classes they remembered this observant lady with a special hatred: they accorded her an honor hitherto reserved for unpopular Egyptian pharaohs. They violated her tomb and pitched out her bones.

Not all of the aristocratic ladies were as sensitive as Madame. The noble ladies and gentlemen fumed at their boredom, enforced by the presence of Mazarin's barbarians. When the troops went into winter quarters the ladies and gentlemen relaxed, donned their most expensive costumes and resumed their customary pleasures, stimulated by new and exciting war games.

The games, inspired by the "Frondes," were sometimes quite macabre. The years of the Thirty Years' War were almost civilized compared to the immediately succeeding years. While the ladies and gentlemen were absorbed in their pastimes, hordes of starving beggars and cripples, fleeing from plundering mobs of undisciplined soldiers, swarmed over the country seeking refuge. Religious orders offered what relief they could, which was little enough. The hospitals had to turn away most of the suffering and indigent. This was not a flood of misery, it was a deluge. There was no effective command to control the private armies ravaging the land. In desperation the fleeing peasants implored the nuns and the monks to accept and care for their livestock rather than let it fall into the hands of the military. As the monasteries had no room for even a small part of the refugees, the peasants lived with their animals. The churches took care of the household utensils and furniture, the rags, and as much of the crops as the peasants could salvage before the advancing soldiers. But for all their good will the churches and monasteries could not accommodate the distracted human beings. Some hid in the woods, others holed up in caves and so became prisoners of war—mediaeval chivalry was not yet completely dead, though decaying. Not to be burdened with starving prisoners who could barely walk, the soldiers lit fires at the entrances of the caves, thus asphyxiating the helpless inmates. Though not so intended, this was a merciful end for human beings who had passed the limit of starvation and were gnawing at what strings of tendons remained on their own hands and arms.

Civil war and anarchy supplied their quotas of dead. Sanitation was nonexistent. The corpses were left unburied where they had fallen. One battle contributed 1500. The cemeteries could not absorb the supply; at Rouen 17,000 died in a year, at Dreux (1651) 500 out of 4000 died. While the crawling sick burrowed into the heaps of decomposed carrion for anything edible, elegantly gowned ladies—"soup butchers"—dispensed watery sustenance to such of the peasants as could still walk. There were not many to profit by this bounty. Plague had outstripped most of the military in their marches to relatively undevastated territory.

And where was the Sun King shedding the light of his countenance during all this glorious campaign? Most of the time snugly shacked up with one or another of his numerous lady friends. He had wisely delegated important details of the military operations to the capable and diligent Mazarin.

We must now glance at taxation under Louis. His financial profligacy demanded tax upon tax upon tax down to the ultimate contributor to the king's welfare—the peasants.

The tax structure of France in Louis' time was so complicated that little more than a mere mention of a few of the more profitable extortions can be attempted. There is an almost ludicrous similarity between certain of the French

285

schemes for raising money and their equivalents in the United States today. Sometimes the parallels are so close that we are tempted to believe that time has been flowing backward since about 1600 and thus that the French copied the Americans. But this of course is physically, if not morally, impossible according to the Second Law of Relativity. Who invented withholding taxes and capital gains taxes? The French of the seventeenth century. Did the Americans conceive the ingenious obscenity of mating these two—withholding and capital gains taxes—to generate a double squeeze? No; this was a typically French chef-d'oeuvre. It worked admirably in France. The peasants could not turn over in bed, or anywhere else, without feeling the squeeze.

Two main taxes, the taille and the gabelle, outranked the others in the scope and effectiveness of their exactions. What we should call excise taxes, individually did not amount to very much; in their multitude *cooperatively* they were as distressing as a swarm of fleas on a dog's back. There was one excise tax the French did not pay, for obvious reasons—that on cigarettes. With that happy exception we pass on to extortions shared by nearly everybody—the nobility and the clergy were relieved from the more crushing burdens.

First, the taille, or tallage. Our closest equivalent in the United States is the general property tax levied on land and the buildings on it, also on what we should call income. There was an important distinction, however, and in our favor: our taxes are fixed by certain legally constituted and responsible bodies and are not, as the French taxes were, fluid and left to the tax *collectors* to estimate by their own arbitrary whim. Whatever they could squeeze was legitimate. The outcome of this policy could have been foreseen by anyone but a low-grade moron. Instead of bleeding only the scrawny peasants, the collectors cut their own fat gullets: the peasants made no

286

attempt to improve their methods of cultivation; increased yield meant a comparable increase in value and hence in taxable property; the sensible course for the peasants was to sit down, do nothing, and let the land lapse into sterility. This implied an accompanying deterioration of buildings and equipment; both rotted. As long as they could exist, at no matter how depressed a standard of living, the peasants let the farms go barren. Like the land, they were starving. The natural ambition of normal human beings expired in apathy and indifference. It will be noted presently how the French equivalent of our withholding tax produced similar results.

Louis was disconcerted at the lack of revenue from many sections. But he was not dismayed. New taxes would restore his prosperity. The taille had flopped, temporarily; the gabelle could be increased practically indefinitely; the day would be saved by the gabelle.

It is generally admitted that salt is a necessity for the health of man and beast. Wherever there is a necessity, there also is a bottomless source of collectible taxes. The gabelle, the salt tax, should be doubled, even trebled or quadrupled to make up the deficiency caused by the collapse of agriculture. Probably it was Mazarin who proposed this particular squeeze. It accomplished what was wanted, also something that was not wanted, a revolt of the peasants—the taxpayers. (A recent American parallel to the gabelle was the successive increases in postal rates. By a singular coincidence it was noticed first that a deficit of many million dollars could thus be made up by the Post Office Department, as a servant of the people, in a futile attempt to "balance the budget." This piddling preliminary subsidy, later substantially increased, went down the drain almost unnoticed by the taxpayers; a few grumbled at the increased cost required for mailing a letter. That was all. There was no subversion as in France over the gabelle.)

Solicitous for the health of his taxpayers, Louis, at the suggestion of his expert squeezers, including Mazarin, decreed that every family should buy annually a fixed amount of salt, the amount being proportional to the size of the family. To ensure that the buyers actually bought the forced amount of salt, whether they needed it or not, the collecting of the tax was "farmed out" to independent and irresponsible agents, and the peasants had to buy salt from these leeches alone. Only a saint or a blind man could have overlooked the obvious invitation to gouge and profiteer. The *"gabelous,"* as the collectors were called, had the right of entry and search at any hour of the day or night to see that the regulations were being observed. This constant prying exasperated the peasants to the point of rebellion, but they bought salt and paid till they went broke. The tax farmers squeezed what they could by any means that occurred to them, including false testimony and perjury before corrupt magistrates.

Fermat noted and tried to rectify this abuse. The charges against the peasants were allegations of trivial breaches of the rules. Bribes or exorbitant fines might get some off. If the accused could not pay he went to a dungeon. By a sort of inverted justice, the corruption and greed of the *gabelous* cut the king's share of the tax to less than a fourth; the *gabelous* pocketed the rest. Louis, like the proverbial fool he was, was notoriously easy to part from his money; his agents and collectors robbed him of all but a small fraction of the take.

The French anticipation of at least some features of the American withholding tax can now be very briefly noted. There was no reliable index of what the taille would be in a particular year. The peasant had to guess what the tax collectors might say he owed according to the amount of the estimated crop. If he guessed wrong he was penalized either way: if the crop fell short of the estimate, the tax collectors took the lot; if the estimate exceeded the actual crop, the "excess profit" was confiscated and the peasant heavily fined for his poor guess. So it was impossible to win. As we have noted, the peasants very sensibly did not attempt to cultivate the land; an increase in the crop meant an increase in the tax; barren fields meant no crop and therefore no tax. It was better to starve than to compound the felony of royal robbery. This referred to the civil authority. But the taille was partitioned in several ways, each of which was felt and resented by the taxpayers. The Church, for instance, took a sizable cut of the total as its share of the loot, by tithing and other respectable survivals of feudalism.

What about the actual collectors of the taille? They could not go about their business unattended. Troops with orders to *enforce by all means in their power* the extortion of what was owed to Church and state accompanied the collectors on their forays. Resistance meant death, or the ignominy of slowly starving and rotting in the packed dungeons of the collectors. Rather than submit, scores of debtors exercised their last human right and committed suicide.

One detail that might be inferred from the narrative so far was the wholesale application of torture, accepted without question, as an instrument in the administration of the law. I refrain from attempting to describe the details—not that they are indescribable, for they have often been depicted, but because an account of it would be superfluous and perhaps tame after our own memories of the 1930s and 1940s.

In addition to confiscatory taxation and exhausting wars, the inordinate extravagance of royalty helped to beggar France. That impressive shrine of tourists, the Palace of Versailles, said now by Texans to be unfit for stabling cattle, was costly in more than money and taxes. By some bureaucrat's blunder the Palace was planted in a fever-stricken swamp. It cost the lives by "marsh fever" (malaria?) of thousands of workmen. From first to last the lives of more than 30,000 laborers were thrown away. In money 700,000 livres went down the chute. If human life was cheap, money wasn't, and the unrestrained squandering demanded new taxes, which the survivors of the costly debacle and their next of kin donated for the king's vanity and glory. The completed work was pompously grandiose and all but uninhabitable by normal human beings. A horde of 10,000 courtiers could be accommodated without jostling in the main building and its appendages.

In all the magnificence and splendor of Versailles there was not the slightest hint of sanitation. But the 10,000 courtiers, like any other human beings, had to heed nature when she called. To obey that imperative summons, they relieved themselves wherever they might be at a particular moment—the stairways were a favorite refuge. The resulting stench was only slightly abated by the costly Parisian perfumes with which the ladies doused their skirts. Neither sex ever bathed or washed their hands; the belles occasionally would give their faces a perfunctory swipe with a dry *mouchoir*. Where everybody stank, few noticed anything unpleasant; *eau de merde* was the universal equivalent of the later eau de cologne.

Other heaps of masonry, less pretentious than Versailles, also helped to drain the treasury and pile up bad debts. The cost of all these magnificences has long since been absorbed and more or less forgotten in the tax bill, so no tourist conscience need be unduly perturbed by the question, "Who paid for it all?" The answer is as plain and as simply true as that to a similar query would have been in the Egypt of Cheops (the Egyptian "Sun King"). If there is any advantage between Egypt and France it is on the side of Egypt. The Egyptians had at least a fairly clean river, the Nile, to flee to and drink from. Had the French sought such a refuge, say in the Revolution, their only "water" would have been the curdling sewage of the Seine. As for Paris the beautiful, it could have done with a thorough job of sanitation:

> A Paris gutter in the good old times,
>> Black and putrescent in its stagnant bed
>> Save where the shamble oozings fringe it red,
> Or scaffold trickles, or nocturnal crimes...

> It holds dropped gold; dead flowers from tropic climes;
>> Gems true and false, by midnight maskers shed;
>> Old pots of rouge; old broken phials that spread
> Vague fumes of musk, with fumes from slums and slimes.

(Though this sounds like Baudelaire, it is only from Lee-Hamiltons's poem about that connoisseur of decadence and decay.)

The streets of Paris in the "good old times" of Louis XIV had neither curbs nor sidewalks. They were churned-up bogs of mud and decomposing garbage in wet weather; in dry they exhaled a stench of corruption that could be smelled two miles from the city. In spite of all these unpleasant messes, elegant ladies, beautifully gowned, managed to get from a cold assignation to a hot one in their coaches through unpoliced and unlighted streets, and only occasionally were their gentlemen escorts, beggarly "nobles of the sword," overmastered. L'amour, toujours l'amour! Such were "the good old times" in Paris.

291

14

The Jurist

> Stern Daughter of the Voice of God!
> O Duty; if that name thou love
> Who art a light to guide, a rod
> To check the erring, and reprove;
> Thou, who are victory and law...

This apostrophe from Wordworth's noble "Ode to Duty" might have been declaimed by Fermat or any of his swarming colleagues dedicated to the parasitic service of the law in seventeenth-century France. The only lack is a declaration that the King of France was God's delegated "rod to check the erring, and reprove" all hungry malcontents who should dispute the divine right of their anointed king to restrict the diet of his dutiful subject to grass. Considering Fermat's France, we might be justified in twisting Burns's verse in his apology to the field mouse whose dwelling the plow had upturned—

> Man's inhumanity to man
> Makes countless thousands mourn,

by putting *humanity* for its opposite. This, of course, has been remarked before.

Of all the revered fetishes adored by a docile humanity, the two most abjectly worshiped in Fermat's day, in both France and England, were duty and loyalty. It does not seem to have occurred to the dutiful and the loyal of France to question the authority of the king—backed by the "victory" of uniformed and predatory ruffians—to tax them bloodless till they stumbled into ditches and starved to death, their mouths stopped with unchewed grass. That particular kind of disloyal and undutiful doubt was not to come till about a century after Fermat was dead, when it was sharpened by the guillotine. All his mature life Fermat did his whole unquestioning duty to his just and tyrannical God and his stupid and tax-greedy king. Only once did he seem to hesitate for a moment. But it was a passing infidelity. He loyally and dutifully repressed his impulse to ameliorate the bestial degradation and intolerable misery of his fellow subjects under God and the king. He dutifully and loyally turned his back on the helpless. They were outside the reach of help. Their one hope was to help themselves by the extreme of brutality.

To repeat the cliché, Fermat was of his times, as we are of ours, and censure is not only irrelevant but impertinent, perhaps especially from us. The facts reported later are their own justification and Fermat's. In his day it was not only lawful but dutiful for a judge to tax the peasantry into squalor and death by starvation. Nor was it unusual for a conscientious judge to condemn a man to "death by fire"—to be burned alive—for reasons of ecclesiastical polity or financial expediency. The Kingdom of God on earth had to be sustained no less vigorously than the bailiwick of the reigning King of France. That the two necessities frequently coincided during Fermat's service as a jurist was to the advantage of both. No more dispassionate judge than Fermat could have been found in all the

293

provinces of France. He discharged his duties long and honorably.

Writing to Martin Cureau de la Chambre on August 18, 1648, Fermat says he "has not kept him posted till now on public affairs because the actual consequences of a decree handed down by the Parliament are perhaps unknown to Monseigneur." He continues that he has written a memorandum in which De la Chambre will find what is necessary for understanding the situation.

I "expose" this with the assurance I have in your prudence and the confidence that you love me. It need see only what light you wish... I do not claim to make this appear as anything more than my zeal for the service of the King and my respect for the wishes of Monseigneur. If this memorandum should not serve for that, at least excuse my faults and do me the grace to keep them hidden, and to believe me always, Sir, your very humble and very obedient servant, Fermat.

It is the memorandum attached to this note which is of interest here, as showing the two sides of Fermat's character, his humanitarian sympathies, and his uncompromising sense of duty to his office and his loyalty to his king. The gist of it follows.

The decree handed down by the Parliament of Toulouse forbids the levying of tallage by armed force and by billeting of soldiers. This was necessary in the present juncture because it apparently was the only remedy for maintaining calm in the province of Guinne, which is dependent on this jurisdiction.

The rumor that spread through all the towns that the King was going to cancel arrears [in the payment of taxes] and grant a considerable reduction of the current tallage, caused the people to bear these harsh orders for billeting [of soldiers who robbed them] with so much impatience that on all sides there were plots and mob uprisings against the brigades, and such considerable rebellions that without doubt they would have hatched greater uprisings if the Parliament had not, by its decree, suspended these violent orders [about billeting] which are contrary to the ordinances and contrary also to very humanity, if it must be said.

"Ah," the hasty reader may exclaim, as did a French friend who read the whole memorandum, "Fermat was on the side of the angels." His sympathies might have led him there; his loyalty put him squarely on the side of the king with a brilliant suggestion for a trick of legal chicanery which would give the king his money and keep the people starving quietly a little longer. Passing from humanity to taxable human beings, Fermat reports that diligence in tax collecting has by no means been relaxed. Further, illegal devices by which the tax collectors extorted even more from the peasants than the king exacted were effectively outlawed. "I was the first to gain some knowledge of these crooked ways, and I suggested to the High Court [Grande Chambre] the decree issued on this matter." He says the decree was necessary because "poverty is so general and so great." But if the taxpayers could not be further squeezed, how was the king to continue luxuriating in the style to which he had been accustomed? Fermat the jurist gives way here to Fermat the mathematician. His solution of the king's problem is as simple as it is masterly.

The levies must be hurried up and the King be promptly given so just and so necessary help. It seems to me the most plausible and easiest expedient is to have a declaration from the King which would carry permission to all the communities to borrow the necessary funds to meet the current taxes (tallages), and which would declare the sums borrowed for this effect privileged to meet

295

all debts of the said communities in payment of current charges. It is very probable that all the money of the province would end up there, for the frequency of bankruptcies is the reason why those who have money prefer to keep it rather than to risk it. This declaration, signed by the registrar of Parliament, would be a full assurance for the covenantees; and if the King were to grant some delay in advance, all the communities would rush in a crowd to borrow the necessary funds and pay them as quickly as possible to the receivers. One could even enjoin Parliament to send commissioners to all the towns to facilitate the said payments; and if His Majesty should judge that it were important for his well being to make use of this same means to cause the funds to draw in for the year 1649, its execution would not, apparently, be difficult.

This ingenious scheme of Fermat's for raising the dough to keep the king healthy and happy, has some of the marks of a skin game. What an admirable Secretary of the Treasury or Chancellor of the Exchequer in times of "national emergency" Fermat would have made!

Fermat's duties compelled him to serve not only the King, but the higher authority of God as interpreted by the Church. For the incident recounted presently we have the authority of that notorious liar, Sir Kenelm Digby, in a letter of February 6, 1658 to John Wallis. It is a long letter, written in Digby's usual style of humble sycophancy and oily insincerity. But for once his testimony rings true. He has been trying, and had failed, to see Fermat personally in behalf of Wallis. He was properly snubbed.

...I put off writing to you [Wallis], because I thought he [Fermet] would send his reply by the first or the second courier. But, since then, I have had nothing from him but successive excuses, always putting me off to the next time.

The relevant part of the letter follows.

It is true that I had exactly hit the date of the displacement of the judges of Castress to Toulouse, where he [Fermat] is the Supreme Judge to the Sovereign Court of Parliament; and since then he has been occupied with capital cases of great importance, in which he has finished by imposing a sentence that has made a great stir; it concerned the condemnation of a priest, who had abused his functions, to be burned at the stake. This affair has just finished and the execution has followed.[1]

---

[1] It would be interesting to know exactly what the priest had done. Digby does not say. He continues with his grievance that Fermat did not answer his letters. The "affair" of the priest was no excuse for Fermat's impoliteness: "But what might be an excuse for another is not for M. Fermat, who is incredibly sharp and penetrating in everything he undertakes."

15

Aftermath

(A final note by D. H. Lehmer)

I have been asked to append to Professor Bell's very readable history of the Fermat Problem a very short account of the problem since the seventeenth century. For a complete history up to 1920 the mathematically competent reader is referred to the last chapter of Volume 2 of L. E. Dickson's *History of the Theory of Numbers*.

As Professor Bell has explained, Fermat himself gave a beautiful proof by descent of the fact that

$$x^n + y^n = z^n$$

is impossible for $n = 4$ and hence for n any higher power of 2. Hence it may be supposed that n is divisible by some odd prime l. But in this case an nth power is automatically an lth power and so it suffices to consider the equation

$$x^l + y^l = z^l$$

and to prove that this equation is impossible in positive integers x, y, z for every prime l greater than 2. Not only have mathematicians been unsuccessful in proving this for every such l, but no one has been able to handle the problem for an infinity of primes l. Incidentally, the choice of the letter l in this connection is due to the great nineteenth-century German mathematician E. E. Kummer. Although any other letter (except of course x, y, or z) would serve equally well and despite the fact that the letter l is ambiguous on the twentieth-century typewriter, this letter has now become traditional with "professional" workers on the problem.

The eighteenth and early nineteenth centuries produced only minor results: proofs of the theorem for $l = 3, 5, 7$, and 11 by special methods and equivalent restatements of the problem. The first great advance on the problem was made by Kummer in 1847. He wrote the equation in the form:

$$x^l = z^l - y^l$$

and proceeded to factor the right-hand side into factors of the form:

$$z - \epsilon^h y$$

where ϵ is a complex number whose lth power is l. For this approach Kummer had to invent a new system of complex "cyclotomic" integers and to consider the prime factors of such numbers. All went well until he came to the case of $l = 37$. To his surprise and disappointment the cyclotomic integers in this case are not products of primes in only one way. This failure of unique factorization into primes vitiated his method of proof for $l = 37$ and for all other such primes, which he called "irregular." The attempt to restore unique factorization led Kummer to invent "ideal numbers," described by Dickson as "one of the chief scientific triumphs of the last century." This

299

allowed Kummer to dispose of $l = 37$, 59 and 67, the only irregular primes less than 100, and, in general, to obtain quite stringent new conditions on the irregular primes l.

The twentieth century saw an upsurge of interest in the Fermat problem by amateurs because of a prize of 100,000 marks offered by a wealthy German mathematician, P. Wolfskel, in 1909 for a published solution of the problem. The prize money was never awarded and it disappeared in inflation. Thousands of erroneous self-published proofs were submitted but the whole lot failed to advance the problem one inch. Even with the vanishing of the prize money, "proofs" of the great theorem continue to be submitted, though at a reduced rate. The great German number theorist E. Landau is said to have had post cards printed with the following message: "Dear Sir or Madam: Your proof of Fermat's Last Theorem has been received. The first mistake is on page..., line..." He would give these cards to graduate students to be filled in and posted.

The twentieth-century work on the problem has been in the direction of obtaining further criteria from Kummer's analysis and attempting to combine them so as to lead to a more stringent (and hopefully a contradictory) set of conditions on l.

Early in the history of the problem it was recognized that there are two cases. The so-called first case is that in which l does not divide xyz. In the second case l divides x. Contrary to what one might expect, the first case is much more tractable than the second. In fact Professor H. S. Vandiver, the world's leading authority on the problem, says in his *Encyclopaedia Brittanica* article that it is indeed remarkable that the first case is still unsettled. By combining fourteen different conditions D. H. and Emma Lehmer succeeded in 1941 in proving that the first case is true for all primes l less than 253747889. The largest prime l for which the first case is known to be true is the Mersenne prime $2^{3617} - 1$, a number of 1089 digits.

As for the second case, since Kummer has settled the problem when l is regular, we can now consider only the irregular primes. Unfortunately Jensen has proved that there is an infinity of irregular primes. Whether there are infinitely many regular primes is not known. However, there is ample evidence that three primes out of five are regular.

Under the guidance of Vandiver, J. L. Selfridge, C. A. Nicol and the two Lehmers, the electronic computer SWAC has proved that Fermat's theorem is true for l up to 4001. At this point each irregular prime l required about an hour for the machine to make the proof and so the project had to be abandoned as too expensive. It can be proved that, if l is greater than 4001, then z, if it exists, has more than 43255 digits. Until the nonexistence of all possible numbers x, y, z, l, large or small, is definitely proven, or until a set of numbers is found to satisfy the equation, the great problem will still be with us. Progress up to now may be fairly described as fabrication, preparation and comparison of various weapons for assault on the fortress. Direct attack and Trojan horse tactics have thus far failed to take the citadel.

Notes for *The Last Problem*

Underwood Dudley

Chapter 1

Page 18.

Goldbach's conjecture that every even number greater than two is a sum of two primes is almost certainly true. Not only is it easy to write even numbers as sums of primes, it gets easier as the numbers get larger: there is evidence, both numerical and theoretical, that the number of ways of writing $2n$ as a sum of two primes is on the order of $n/(\log n)^2$, a quantity that approaches infinity rapidly with n. However, there has been no proof of this, nor is there any sign that one is coming. The theorem of Vinogradov mentioned in the text was found by using a modification of the Hardy-Littlewood circle method. Density theorems have also been applied which, unlike Vinogradov's result that applied only for "sufficiently large" integers where exactly how large that was either could not be specified or turned out to be very large indeed, give results about every integer, leading to Vaughn's 1976 result that every even integer is a sum of at most six primes. Successive refinements of a third idea, that of using sieve methods, have come the closest to settling the conjecture. In 1923, Rademacher showed that every even integer was the sum of two numbers, each with at most 7 prime factors. Let us abbreviate this as $P_7 + P_7$. In 1932, Estermann improved the result to $P_6 + P_6$. Further improvements followed, including $P_4 + P_4$ (Buchstab, 1940), $P_1 + P_4$ (Wang, 1956), and $P_1 + P_3$ (Buchstab, 1966).

Around 1973, Chen achieved $P_1 + P_2$, just one little factor too many, and there the conjecture stands today.

One reason why proving the conjecture is so difficult is that it is about addition, and prime numbers do not arise from addition: they are the multiplicative building blocks of the integers. When the conjecture is proved, if it ever is, I conjecture that it will be shown that it is true for any sequence of integers sufficiently like the primes—that is, any sequence fairly dense in the sequence of integers, with no very large gaps between its members.

The book *Goldbach Conjecture* [1] contains a collection of original papers on the conjecture, translated into English from Chinese, French, German, or Russian when necessary, going from Brun's sieve (1919) and Hardy and Littlewood's circle method (1922) all the way to Chen's result. The papers can be read only by specialists, but anyone can look at them and read their introductions to get a feel for the history of the problem and how progress has been made. The whole of the book supports the arguments of those who say that mathematics advances, and new mathematics is discovered, because of attempts to solve specific problems.

Reference
[1] Wang Yuan, editor, *Goldbach Conjecture*, World Scientific Publishing, Singapore, 1984.

Chapter 2

Page 43.
When Bell said, "There may be no general, comprehensive theory [of Diophantine equations]. It may be impossible to construct one" he was probably thinking of Hilbert's Tenth

Problem. The tenth of David Hilbert's famous twenty-four problems presented to the mathematical world in 1900 was to find an algorithm that could be applied to any polynomial Diophantine equation to determine if it had solutions in integers. Notice that Hilbert did not go so far as to pose the problem for *all* Diophantine equations. That would clearly be impossible to solve, but Hilbert might have thought that there was a hope for a solution for a restricted class of equations. Hilbert lived in an optimistic time, when it seemed within reach to prove that there were no possible contradictions in Euclidean geometry, that mathematical physics could be axiomatized, and that all of mathematics might be put on a foundation as solid as rock. If all of those things were possible, it might be possible also to do something for all of Diophantine equations. But, alas, Gödel's incompleteness theorem of 1930 showed that there could be no rock, only sand forever and ever, and hope for universal methods faded.

Seventy years after it was proposed, Hilbert's tenth problem was solved: it was proved that there is in fact no universal method. It is natural to wonder how such a result could possibly be established. How can we be sure of the nonexistence of something? We cannot examine all possible methods and perhaps we might miss the right one, either by bad luck or because we were not smart enough to find it. The start of the road towards the proof was in work done in 1952 and 1953 by Julia Robinson and Martin Davis, with the idea not of looking at Diophantine equations but instead at Diophantine sets: sets of numbers that are solutions of Diophantine equations. The proof of the impossibility was reduced to the question of whether a set was computable. ("Computable" has a technical meaning that does not conflict with the usual meaning of the word.) The subject was investigated by many people over the years, notably by Hilary Putnam in the early 1960s, but it was not until 1970 that Yuri Matijasevic was able to complete the proof. The proof (not the hardest part) involves Cantor's diagonal process, which is one reason why it does not allow us to find a specific Diophantine equation that we cannot solve. But, no matter what method we use, there will always be such equations, just as whatever method we use to list the irrational numbers, there will always be some that are not on the list. Combining this with Gödel's theorem, we know that there are polynomial Diophantine equations that have no solutions that we will never be able to prove have no solutions. But we have no examples to point to. Fermat's equation may be one of this class; then again, it may not.

Martin Davis [4] wrote an exposition intended for the general mathematical public that serves, if following the details is too difficult, at least to give the flavor of the solution. There is also a survey of the problem and its positive results [5] that contains seventy-eight references.

Page 52.

It is true that many special cases of unit fraction expansions have been disposed of and that the general case is still outstanding. Nevertheless, work continues: for example, it is now known that $12/n$ can be written as a sum of three unit fractions for all $n > 12241$. Some references to some of the work can be found in [2, 3]. Unit fractions are not serious mathematics, if "serious mathematics" is defined to be what serious mathematicians do. Expressing 1 as a sum of distinct reciprocals is not serious—talks are not given on it, no one gets grants to study it, and promotion, tenure, and salary increases are not based on it—but it is fun enough that [1] lists twenty references.

References

[1] Barbeau, E. J., Expressing one as a sum of distinct reciprocals, *Eureka* (now *Crux Mathematicorum*) 3 #7 (1977) 178–180.

[2] Brenner, J. L., Comment on the solution of problem 346, *Crux Mathematicorum* 10 #9 (1984) 293–294.

[3] Campbell, Paul J., A "practical" approach to Egyptian fractions, *Journal of Recreational Mathematics* 10 (1977–78) 81–86.

[4] Davis, Martin, Hilbert's tenth problem is unsolvable, *American Mathematical Monthly*, 80 (1973) 233–269.

[5] Davis, Martin, Yuri Matijasevic, and Julia Robinson, Diophantine equations; positive aspects of a negative solution, in *Mathematical Developments Arising from Hilbert Problems*, American Mathematical Society, Providence, Rhode Island, 1976, 323–378.

Chapter 3

Page 85.

Bell's assertion that dozens of proofs of the Pythagorean Theorem have been constructed is an understatement, if he meant only a few dozen. The 1968 reissue of Loomis's book [1], first privately printed in 1927, contains *370* proofs of the theorem. Many of them bear strong resemblances to each other, but even so the number of different approaches is remarkable.

Page 91.

While Cauchy did give the first proof of the general polygonal number theorem, he did not find it entirely on his own. He was standing on the shoulders of giants. Legendre had shown in 1798 that every integer is the sum of three triangular numbers, a theorem later independently proved by Gauss. Euler had tried to show for decades that every integer is a sum of four squares and failed; the first proof was by Lagrange in 1772.

Fermat's assertion that he had a general proof was, I think, wrong.

Reference
[1] Loomis, Elisha Scott, *The Pythagorean Proposition*, National Council of Teachers of Mathematics, Washington, D. C., 1968.

Chapter 6

Page 141.

Bell, with the modesty characteristic of mathematicians who assume that anything that they can understand must therefore be comprehensible to everyone, overestimates the ability of twelve-year-olds to find greatest common divisors. They, and many of their elders, would need to see an example of the algorithm for finding greatest common divisors written out:

$$75185353 = 1 \cdot 72829807 + 2355546$$
$$72829807 = 30 \cdot 2355546 + 2163427$$
$$2355546 = 1 \cdot 2163427 + 192119$$
$$2163427 = 11 \cdot 192119 + 50118$$
$$192119 = 3 \cdot 50118 + 41765$$
$$50118 = 1 \cdot 41765 + 8353$$
$$41765 = 5 \cdot 8353.$$

Page 144.

How computers have changed the world! It is no longer true that we might spend months or years to see if a number $2 \cdot 3 \cdot 5 \cdot 7 \cdot 11 \cdots p + 1$ is prime for moderately large p like 9973. Now it takes only seconds or minutes. The largest known prime of that form is now the one with $p = 18523$, a number with 8002 digits. This was discovered by Harvey Dubner [7] who has constructed a special-purpose number theory computer that has found many other large primes, including $1477! + 1$, $1963! - 1$, and $2 \cdot 3 \cdot 5 \cdot 11 \cdots 15877 - 1$. These discoveries do not count as serious mathematics either, but it could be argued that they are at least as interesting and important as many of the theorems that appear in journals and immediately sink without a trace, never to be seen again. Details on the computer can be found in [8].

Page 146.

The problem of whether odd perfect numbers exist remains unsolved. Several researches, notably Peter Hagis, Jr., have found numerous necessary conditions that odd perfect numbers would have to satisfy and which make their existence very unlikely. In particular, the smallest odd perfect number would have to have more than 50 digits. Every odd perfect number has a prime-power divisor greater than 10^{30} [5]. Every odd perfect number has a prime factor greater than 100,000 [3]. The necessary conditions go on and on. There are some references in [10]. The problem is like Goldbach's Conjecture in that multiplication—the divisors of an integer—is combined with addition—summing the divisors to get twice the number —and that is why it is likely to stay unsolved for a long time.

Page 156.

The calculations of the Hillsboro (Ill.) Mathematical Club on the solution of Archimedes' Cattle Problem *have* been checked, and the last two of their left-most digits were wrong. A complete calculation of the 206545 digits in the answer was made as long ago as 1965 but the solution was not printed. It was recalculated by Harry Nelson [14] *and* printed. Nelson used the computation as a test for a new computer and found that the machine did it too quickly to provide a good test: he had to calculate more solutions, some with more than a million digits, to give the machine a proper workout. Before starting to look for curious sequences of digits in the solution, like 123456789 or 31415926535, be warned that it is printed with 1/3-sized digits.

Page 157.

It was in 1958 that the "enthusiastic idiots" calculated 10,000 decimals of π. People with equal enthusiasm extended the computation to 100,000 places in 1961, to 1,000,000 in 1973, and to 16,000,000 in 1983. In 1987, the record was raised to 134,217,000 places by Yasumasa Kanada and, as J. M. and P. B. Borwein say [2], a few more weeks of machine time would have given 2 billion digits. Kanada achieved 201 million digits in 1988, and Gregory and David Chudnovsky ingeniously found 1,011,196,691 digits in 1989. It is easy to sneer at these exercises, as Bell did, but they are indications of the health and vigor of the mathematical organism. The difficulty is that things of this sort tend to be publicized, and they tend to be the *only* mathematical items brought to the attention of the public. Thus misconceptions of what mathematics is and what mathematicians are about become yet more firmly planted in the collective consciousness. It seems that there is really nothing that can be done about this.

For an account of almost everything that an ordinary person would want to know about π, see [4]. If you have difficulty remembering its first few digits there are mnemonics in five languages in [17].

Page 160.

There is no new evidence about when Diophantos lived, but Bell's assignment of him to the first century is no longer the common opinion. Writers of histories of mathematics are now almost unanimous in placing him around 250. One reason for this is that Diophantos wrote in a style that resembled that of the Rhind Papyrus, an Egyptian textbook of mathematics: a sequence of specific problems with specific recipes for their solution. It is reasonable to speculate that Diophantus was a Hellenized Egyptian. If so, it is more likely that he could have flourished in the third century than in the first, since the Greek mathematical tradition was then in a more rapid decline.

Page 164.

In saying that "anyone with a little skill in elementary algebra" can prove that cubes are sums of successive odd numbers, Bell is being modest again, or he is underestimating the skill necessary to see that

$$\sum_{i=1}^{n} \left(n^2 - n + (2i - 1) \right) = n^2 \cdot n - n \cdot n + 2 \cdot \frac{n(n+1)}{2} - n$$

$$= n^3 - n^2 + n^2 + n - n = n^3$$

for $n = 1, 2, \ldots$.

Page 165.

The puzzle about Diophantos's age, which Bell calls "silly" and "puerile", does not deserve such contempt, since I am sure that it would baffle almost everyone, students of mathematics included. Its original form (translated from the Greek) was

God granted him to be a boy for the sixth part of his life, and adding a twelfth part to this, he clothed his cheeks with down; he lit him the light of wedlock after a seventh part, and after five years of marriage he granted him a son. Alas! late-born wretched child: after attaining the measure of half his father's life, chill fate took him. After consoling his grief by this science of numbers for four years he ended his life.

The usual interpretation of this is that if x denotes Diophantus's age at death, then

$$\frac{x}{6} + \frac{x}{12} + \frac{x}{7} + 5 + \frac{x}{2} + 4 = x$$

and this gives $x = 84$.

Page 170.

Bell says that "not even the most powerful machine yet invented could [factor] a really large integer." Of course, the truth of that depends on what "really large" means, but progress in primality testing and factoring has been rapid in recent years and I think that Bell would agree that really large integers are now routinely being dealt with. In 1988, a 100-digit integer was factored. Great strides have been taken since 1644, when Mersenne wrote

To tell if a number of 15 or 20 digits is prime or not, all time would not suffice for the test, whatever use is made of what is already known.

Today, a program costing less than $200 running on an ordinary personal computer can take a 20-digit integer, the first twenty digits of π,

31415926535897932384

and find its prime factors

$$2^5 \cdot 563 \cdot 15647 \cdot 93967 \cdot 1186001$$

in less than three seconds. It is not easy to keep up to date, but Hans Riesel's rich and fascinating book [16] gives an account of some of the algorithms, as does [6].

Page 173.

Bell says that Euler thought that $x^4 + y^4 + z^4 = w^4$ was impossible. Euler thought that you needed at least n nth powers to sum to an nth power, probably on the analogy of the solvability of $x^3 + y^3 + z^3 = w^3$ ($3^3 + 4^3 + 5^3 = 6^3$; who could overlook *that*?) and the unsolvability of $x^3 + y^3 = z^3$. Euler was mistaken, but it was not until 1966 that Lander and Parkin [11] found a counterexample. It was not a counterexample for exponent 4, but for exponent 5:

$$27^5 + 84^5 + 110^5 + 133^5 = 144^5.$$

Work proceeded on exponent 4, and it was shown that if there were any solutions, then w had to be at least 220,000 [12]. It often happens on problems like these that the lower bound, 220,000 in this case, gets extended further and further upwards, to the millions and billions and beyond, but never to infinity, nor is any solution found. But here is an exception: in 1988, Elkies [9] found that there were infinitely many solutions for exponent 4, with the smallest being

$$95800^4 + 217519^4 + 414560^4 = 422481^4.$$

Page 174.

Bell says that the reader who remembers some trigonometry "will observe" the similarity of the formula for multiplying two sums of two squares to get a third to the addition formulas for sines and cosines. Maybe so and maybe not. There are those who maintain that some mathematical writers' habit, seemingly because of their humility, of overestimating the knowledge and ability of their readers is only a pose and their motive is really to maintain their superiority. At any rate, the analogy with

$$\cos(a + b)\cos(a - b) = \tfrac{1}{2}\cos 2a + \tfrac{1}{2}\cos 2b,$$

if that is what Bell intended, is not all that close.

Page 178.

There are indeed many more pairs of amicable numbers. The next smallest pair after 220, 284 is

$$1184 = 2^5 \cdot 37 \quad \text{and} \quad 1210 = 2 \cdot 5 \cdot 11^2.$$

Euler once took the trouble to find some 60 pairs, and more and more have accumulated over the years. The subject is summarized, up to 1972, in [13]. There is a conjecture that there are infinitely many amicable pairs, and we may be on the brink of seeing it verified. Recently, H. J. J. te Riele and others have made progress in getting new amicable pairs from old [1, 15], though no infinite chain has yet been found.

References

[1] Bohro, W. and H. Hoffman, Breeding amicable numbers in abundance, *Mathematics of Computation* 46 (1986) 281–293.

[2] Borwein, J. M. and P. B. Borwein, Ramanujan and pi, *Scientific American* February 1988 112–117.

[3] Brandstein, Michael S., Abstract 82T-10-173, *Abstracts* of papers presented to the American Mathematical Society 3 (1982) #2 190.

[4] Castellanos, Dario, The ubiquitous pi, *Mathematics Magazine* 61 (1988) #2 67–98, #3 148–163.

[5] Cohen, Graeme L., On the largest component of an odd perfect number, *Journal of the Australian Mathematical Society* series A 42 (1987) #2 280–286.

[6] Dixon, J. D., Factorizations and primality tests, *American Mathematical Monthly* 91 (1984) 333–352.

[7] Dubner, Harvey, A new primorial prime, *Journal of Recreational Mathematics* 21 (1989) #4 276.

[8] _____ and Robert Dubner, The development of a powerful, low-cost computer for number theory applications, *Journal of Recreational Mathematics* 18 (1985–86) 81–86.

[9] Elkies, Noam D., On $A^4 + B^4 + C^4 = D^4$, *Mathematics of Computation* 51 (1988) #184 825–835.

[10] Laatsch, Richard, Measuring the abundancy of integers, *Mathematics Magazine* 59 (1986) 84–92.

[11] Lander, L. J. and T. R. Parkin, *Bulletin of the American Mathematical Society* 72 (1966) 1079.

313

[12] _____ and J. L. Selfridge, A survey of equal sums of like powers, *Mathematics of Computation*, 21 (1967) 446–453.

[13] Madachy, J. S. and E. J. Lee, The history and discovery of amicable numbers, *Journal of Recreational Mathematics* 5 (1972) 77–93, 153–173, 231–249.

[14] Nelson, Harry, A solution to Archimedes' cattle problem, *Journal of Recreational Mathematics* 13 (1980–81) 162–176.

[15] te Riele, H. J. J., On generating amicable pairs from given amicable pairs, *Mathematics of Computation* 42 (1984) 219–223.

[16] Riesel, Hans, *Prime Numbers and Computer Methods for Factorization*, Birkhauser, Boston, 1985.

[17] Sauvé, Leo, A piece of pi, *Eureka* (now *Crux Mathematicorum*) 1 (1975) #8 81–82.

Chapter 7

Page 192.

The problem whose answer is 380 is isomorphic to the problem of Diophantos's age: in both of them the answer is the least common multiple of the denominators of the fractions. Textbooks never change!

Chapter 8

Page 209.

Bachet's *Problèmes* was the first of many books on recreational mathematics. Their existence is a tribute to the wide and continuing appeal of mathematics. There are no books on recreational metallurgy or recreational sociology that I know of. Recreational mathematics flourished as never before towards the end of the nineteenth century—think of Lucas, Loyd, Rouse Ball, and Dudeney—and continues to thrive today, though probably not for so wide an audience as 100 years ago. The name of Martin Gardner stands out far above all others. In addition to his books and the *Journal of Recreational Mathematics*, Schaaf's bibliographies [1] give an indication of the richness of the field.

Reference

[1] Schaaf, William L., *A Bibliography of Recreational Mathematics*, four volumes, National Council of Teachers of Mathematics, Washington, D. C., 1955, 1970, 1973, 1978.

Chapter 10

Page 231.

Bell carries on the tradition of making mistakes when listing Mersenne primes. When he says, "$p = 61, 81, 107$, not in his list, do yield primes" he errs: $2^{81} - 1$ is divisible by 7. The misprint should be corrected to "89."

Page 232.

More Mersenne primes have been found since Bell wrote. The next largest one after $2^{2281} - 1$ is $2^{3217} - 1$ which took 5.5 hours of computer time to find in 1957. There are now 31 Mersenne primes known, though when you read this there may be more because there are probably at this moment computers searching for them in the milliseconds they have to spare from other work. The list of exponents is

$$2, 3, 5, 7, 13, 17, 19, 31, 61, 89, 107, 127, 521, 607,$$

$$1279, 2203, 2281, 3217, 4253, 4423, 9689, 9941, 11213,$$

$$19937, 21701, 23209, 44497, 86243, 110503, 132049, \text{ and } 216091.$$

In 1983, when the largest exponent known was 86243, Schroeder [2] predicted that the next largest exponent would be in the neighborhood of 130000. Since the next Mersenne prime found was $2^{132049} - 1$, this seemed to a candidate for the best prediction of 1983. However, in 1988 the Mersenne prime with exponent 110503 was found, demonstrating once again the risk involved in making predictions.

315

Page 232.

Progress has been made in factoring the Fermat numbers

$$F_n = 2^{2^n} + 1.$$

They are prime for $n = 1, 2, 3, 4$. F_5 through F_8 are completely factored. F_8 is the product of the primes 1238926361552897 and

93461639715357797776916355819960689658405123754163818858 0280321.

(The Fermat numbers get very big very fast.) In 1989, it was announced [1] that F_{11} had been factored. Also, factors are known for F_9 through F_{20} except for F_{14}, which is known to be composite but for which no factor has yet been found. F_{20} was shown to be composite in 1988 [3], and now the first Fermat number whose status is uncertain is F_{22}, a formidable integer with approximately 1,262,611 digits. The largest Fermat number for which a factor is known is F_{23471}, an even more formidable integer with approximately $3.6 \cdot 10^{7066}$ digits; it is divisible by $5 \cdot 2^{23473} + 1$. That information comes from Hans Riesel's book mentioned in the notes to Chapter 6.

References
[1] Brent, Richard P., Factorization of the eleventh Fermat number, *Abstracts* of papers presented to the American Mathematical Society 10 (1989) #2 176.

[2] Schroeder, M. R., Where is the next Mersenne prime hiding?, *Mathematical Intelligencer* 5 (1983) 31–33.

[3] Young, Jeff and Duncan A. Buell, The twentieth Fermat number is composite, *Mathematics of Computation* 50 (1988) 261–263.

Chapter 11

Page 237.

Sometimes Bell's judgments on people are a bit unfair, or at least not in keeping with common opinion—Descartes is not thought to be as stupid as Bell thought he was—but he seems to have gotten Wallis right. John Aubrey wrote in his *Brief Lives*

Dr. Wallis, (a most ill-natured man, an egregious liar and backbiter, a flatterer and fawner on my Lord Brouckner and his Miss, that my Lord may keep up his reputation).

Chapter 15

Page 305.

The first case of $x^p + y^p = z^p$ is the case where none of x, y, and z is divisible by p. Bell notes that it was known to be impossible for p up to 253,747,889. That has changed. In 1988 there were announcements of improvement to 714,591,416,091,389 [3] and then to 156,442,236,847,241,729 [8]. In another direction, it was proved in 1985 that the first case is true for infinitely many primes; details are given in [4].

Page 306.

The exponent up to which Fermat's theorem is known to be true was 4001 when Bell wrote. In 1976 it was raised to 125,000 [11] and in 1987 to 150,000 [9].

In addition, noncomputational work on the theorem continues, and progress is being made. Progress is being made in *some* non-computational work, that is; hundreds of mathematical amateurs each year attack the problem and tens of them (at least) convince themselves that they have succeeded in solving it when in fact they have not. Their manuscripts are sometimes looked at by mathematicians who sometimes point out where the mistake is and the authors sometimes, but not often, agree that there is an error. (One Fermat's theorem prover wrote his proof in the margin of a book, a nice touch and one that made the error especially easy to spot.) Sometimes, those who think that they have succeeded are not altogether amateurs, for amateurs would not think of inserting announcements of their proofs (one of which was "elementary, zigzag, and truly wonderful") into the pages of an abstracting journal of the American Mathematical Society [12], [13]. It is very unlikely that an amateur will make any significant contribution to work on the theorem.

It is also unlikely that a professional picked at random will make a significant contribution either, but the odds against are not as long, as Gerd Falting's 1983 proof of the Mordell conjecture showed. L. J. Mordell was one of this century's leading contributors to the theory of Diophantine equations, and in 1922 he made a conjecture about them that had implications for Fermat's theorem. His conjecture was about points on curves with coordinates that are rational numbers. If you divide Fermat's equation by z^n, it is $x^n + y^n = 1$ (where $X = x/z$ and $Y = y/z$), the equation of a curve in the plane, and Fermat's theorem is that the only rational points on it are those with one coordinate zero. Mordell's conjecture was that any curve whose equation was an irreducible polynomial with genus at least 2 had only finitely many rational points on it. ("Genus" is a technical term, but the conjecture includes Fermat's equation for $n \geq 4$ because its genus is $(n-1)(n-2)/2$.) The proof of the conjecture was difficult and used many ideas (see [1]) and it was unexpected: an expert writing in 1979 said, "For the moment a proof of Mordell's conjecture seems, with good reason, very remote."

So, we now know that for each $n \geq 3$, $x^n + y^n = z^n$ has at most finitely many solutions. Another consequence of Falting's proof is that Fermat's theorem is true for "almost all" exponents—that is, the percentage of exponents for which it is false is zero. This does not say that the theorem can be false for only finitely many exponents—the percentage of integers that are prime is zero, but there are infinitely many primes—but it insures that the exceptions, if any, are rare and get rarer as n gets larger. It is known that the theorem is true for Mersenne primes, so there are no solutions of $x^p + y^p = z^p$ for the largest known prime. It is known that in any counterexample, x must have at least 1,800,000 digits. Many things are known now that were not known only a few years ago [10]. In 1988 a Japanese mathematician announced that he had proved that there were no solutions, and for a time there was hope that at last we knew it all, and the last problem was solved. However, the proof was found to be incomplete. General opinion now seems to be that the conjecture is true and that we may be on the verge of a proof: see [7] for reasons for the conclusion that "Fermat's last theorem finally seems to rest on reasonably firm ground." This contrasts with Harold Edwards' 1977 statement in his excellent book, *Fermat's Last Theorem* [2], "There seems to me to be no reason at all to assume that Fermat's Last Theorem is true." True or false? Will we live to know?

There are two more references that should be included [5, 6].

References

[1] Bloch, Spencer, The proof of the Mordell conjecture, *Mathematical Intelligencer* 6 (1984) #2 41–47.

[2] Edwards, Harold, *Fermat's Last Theorem*, Springer, New York and Heidelberg, 1977.

[3] Granville, Andrew and Michael Monagan, *Maple Newsletter*, 2 (1988) 10–11.

[4] Heath-Brown, D. R., The first case of Fermat's Last Theorem, *Mathematical Intelligencer* 7 (1985) #4 40–47, 55.

[5] Ribenboim, Paulo, *13 Lectures on Fermat's Last Theorem*, Springer, New York and Heidelberg, 1979.

[6] _____, 1093, *Mathematical Intelligencer* 5 (1983) 28–34.

[7] *Science News* 131 (1987) 397.

[8] Tanner, Jonathan W. and Samuel S. Wagstaff, Jr., A new bound for the first case of Fermat's Last Theorem, *Abstracts* of papers presented to the American Mathematical Society 9 (1988) #6 467.

[9] _____, New congruences for the Bernouli numbers, *Mathematics of Computation* 48 (1987) 341–350.

[10] Wagon, Stan, The evidence. Fermat's last theorem, *Mathematical Intelligencer* 8 (1986) #1 59–61.

[11] Wagstaff, S. S., Jr., The irregular primes to 125000, *Mathematics of Computation* 32 (1978) 583–591.

[12] *Abstracts* of papers presented to the American Mathematical Society 6 (1985) #1 17.

[13] *Abstracts* of papers presented to the American Mathematical Society 8 (1987) #6 410.

Index

E.T. Bell was born in Scotland in 1883, came to this country in 1902, and spent his professional life at the University of Washington and the California Institute of Technology. Besides publishing more than 250 mathematical papers, he was the author of many books on mathematics (*Mathematics: Queen and Servant of Science* is part of the Spectrum series) and also under a pseudonym, some fifteen science-fiction novels. He died in 1960.

What Eric Temple Bell calls the last problem is the problem of showing that Pierre Fermat was not mistaken when he wrote in the margin of a book, almost 350 years ago, that $x^n + y^n = z^n$ has no solution in positive integers when $n \geq 3$. The original text of *The Last Problem* traced the problem from Babylonia in 2000 B.C. to seventeenth-century France. Underwood Dudley, professor of mathematics at DePauw University, has long had an interest in attempts to prove Fermat's last theorem. His notes help to bring us up-to-date on recent attempts to solve the problem.

What T.A.A. Broadbent said about Bell's work applies to *The Last Problem*:

His style is clear and exuberant, his opinions, whether we agree with them or not, are expressed forcefully, often with humor and a little gentle malice. He was no uncritical hero-worshipper, being as quick to mark the opportunity lost as the ground gained, so that from his books we get a vision of mathematics as a high activity of the questing human mind, often fallible, but always pressing on the neverending search for mathematical truth.

This is a rich and varied, wide-ranging book, written with force and vigor by someone with a distinctive style and point of view. It will provide hours of enjoyable reading for anyone interested in mathematics.

Cover Design by Barbieri & Green. Illustration by Max-Karl Winkler